Substrate Noise Coupling in Analog/RF Circuits

For a listing of recent titles in the *Artech House Microwave Library,* turn to the back of this book.

Substrate Noise Coupling in Analog/RF Circuits

Stephane Bronckers
Geert Van der Plas
Gerd Vandersteen
Yves Rolain

ARTECH
HOUSE

BOSTON | LONDON
artechhouse.com

Library of Congress Cataloging-in-Publication Data
A catalog record for this book is available from the U.S. Library of Congress.

British Library Cataloguing in Publication Data
A catalogue record for this book is available from the British Library.

ISBN-13 978-1-59693-271-5

Cover design by Igor Valdman

© 2010 ARTECH HOUSE
685 Canton Street
Norwood, MA 02062

1006164822

10 9 8 7 6 5 4 3 2 1

*To Karen and my family
for their love, trust,
and constant support*

Contents

Foreword   **xiii**

Preface   **xv**

Acknowledgments   **xvii**

Chapter 1 Introduction   1
    1.1 Introduction and Motivation   1
    1.2 Book overview   4
    References   5

Chapter 2 Substrate Noise Propagation   7
    2.1 Introduction   7
    2.2 Modeling the Substrate   8
        2.2.1 Analytical Resistance Calculation Between Two Contacts   9
        2.2.2 Finite Difference Method   13
        2.2.3 Finite Element Method   17
    2.3 The Substrate Modeled with FDM   19
        2.3.1 Experimental Description   19
        2.3.2 Analysis of the Substrate Noise Propagation   20
        2.3.3 Conclusions   27
    2.4 The Substrate as a Finite Element Model   28
        2.4.1 Simulation Methodology   28
        2.4.2 Dealing with N-Doped Regions   30

	2.4.3	Simulation Setup for the Test Structure	33
	2.4.4	Comparison	33
	2.4.5	Conclusions	33
2.5	Conclusions		34
References			36

Chapter 3	Passive Isolation Structures		39
	3.1	Introduction	39
	3.2	Overview and Description of the Different Types of Passive Isolation Structures	40
		3.2.1 The Template Layout	42
		3.2.2 Integrating the Different Types of Guard Rings	43
		3.2.3 Simulation Setup	44
	3.3	Prediction and Understanding of Guard Rings	44
		3.3.1 Reference Structure	45
		3.3.2 P-Well Block Isolation	47
		3.3.3 N-Well Isolation	50
		3.3.4 P^+ Guard Ring Shielding	52
		3.3.5 Triple Well Shielding	56
		3.3.6 Comparison and Conclusion	60
	3.4	Design of an Efficient P^+ Guard Ring	65
		3.4.1 Impedance of the Ground Interconnect	65
		3.4.2 Width of the P^+ Guard Ring	67
		3.4.3 Distance to the Victim	70
		3.4.4 Guidelines for Good P^+ Guard Ring Design	75
	3.5	Conclusions	76
	References		77

Chapter 4	Noise Coupling in Active Devices		79
	4.1	Introduction	79
	4.2	Substrate Noise Impact on Analog Design	80
	4.3	Impact Simulation Methodology	82
		4.3.1 EM Simulation	83
		4.3.2 Circuit Simulation	86
	4.4	Transistor Test Bench	86
		4.4.1 Description of the Transistor Under Test	87
		4.4.2 Modeling the Transistor Test Bench	87
		4.4.3 Experimental Validation	92
	4.5	Substrate Noise Coupling Mechanisms in a Transistor	92

	4.5.1	Analyzing the Different Coupling Mechanisms in a Transistor	94
	4.5.2	Description and Measurement of the Device Under Test	97
	4.5.3	Modeling Different Substrate Noise Coupling Mechanisms	100
	4.5.4	Quantifying the Different Substrate Noise Coupling Mechanisms	104
	4.5.5	Experimental Validation of the Substrate Noise Coupling Mechanisms	106
4.6	Conclusions		108
References			109
Chapter 5	Measuring the Coupling Mechanisms in Analog/RF Circuits		111
5.1	Introduction		111
5.2	Measurement-Based Identification of the Dominant Substrate Noise Coupling Mechanisms		114
	5.2.1	Measurement of the Different Spurs	115
	5.2.2	Sensitivity Functions	116
	5.2.3	Determining the Influence of the PCB	118
5.3	Example: 900 MHz LC-VCO		119
	5.3.1	Description of the LC-VCO	119
	5.3.2	Substrate Sensitivity Measurements	120
	5.3.3	Revealing the Dominant Coupling Mechanism for FM Spurs	123
	5.3.4	Revealing the Dominant Coupling Mechanism for AM Spurs	128
	5.3.5	Influence of the PCB Decoupling Capacitors on the Substrate Noise Impact	130
	5.3.6	Conclusions	132
5.4	Study of the Coupling Mechanisms Between a Power Amplifier and an LC-VCO		133
	5.4.1	Description of the Design of the PPA and the LC-VCO	135
	5.4.2	Coupling Mechanisms Between the PPA and the LC-VCO	137
	5.4.3	Measuring the Dominant Coupling Mechanisms	142
	5.4.4	Conclusions	146

5.5   Conclusions   148
References   148

Chapter 6 The Prediction of the Impact of Substrate Noise
           on Analog/RF Circuits   151
6.1   Introduction   151
6.2   The Substrate Modeled with FDM   152
      6.2.1   Impact Simulation Methodology   152
      6.2.2   Prediction of the Impact of Substrate Noise from
              DC Up to LO Frequency   156
      6.2.3   Experimental Validation of the Simulation
              Methodology   160
      6.2.4   Conclusions   162
6.3   Substrate Modeled by the FEM Method   163
      6.3.1   Impact Simulation Methodology   163
      6.3.2   Prediction of the Impact of Substrate Noise   166
      6.3.3   Verification with Measurements   170
      6.3.4   Conclusions   172
6.4   Techniques to Reduce Substrate Noise Coupling   173
      6.4.1   Layout Techniques to Reduce the Substrate Noise
              Coupling   174
      6.4.2   Circuit Techniques to Reduce the Substrate Noise
              Coupling   175
      6.4.3   3D Stacking as a Solution to Substrate Noise Issues 179
      6.4.4   Separated Analog/Digital Ground   185
      6.4.5   Shared Analog/Digital Ground   186
      6.4.6   Experimental Validation   186
      6.4.7   Conclusions   190
6.5   Conclusions   191
References   192

Chapter 7 Noise Coupling in Analog/RF Systems   195
7.1   Introduction   195
7.2   Impact Simulation Methodology   196
      7.2.1   EM Simulation   198
      7.2.2   Parasitic Extraction   200
      7.2.3   Circuit Simulation   200
7.3   Analyzing the Impact of Substrate Noise
      in Analog/RF Systems   201

	7.3.1	Analysis of the Propagation of Substrate Noise	202
	7.3.2	Analyzing the Substrate Noise Coupling	202
7.4	Substrate Noise Impact on a 48–53 GHz LC-VCO		203
	7.4.1	Description of the LC-VCO	203
	7.4.2	Simulation Setup	203
	7.4.3	Conclusion	207
7.5	Impact of Substrate Noise on a DC to 5 GHz Wideband Receiver		208
	7.5.1	Description of the Wideband Receiver	208
	7.5.2	Simulation Setup	211
	7.5.3	Parasitic Extraction	214
	7.5.4	Circuit Simulation	214
	7.5.5	Revealing the Dominant Coupling Mechanism	215
	7.5.6	Experimental Verification	220
7.6	Conclusions and Discussion		222
	7.6.1	Conclusions	222
	7.6.2	Discussion	223
	References		224

Appendix A: Narrowband Frequency Modulation of LC-Tank VCOs   **227**

Appendix B: Port Conditions   **231**

List of Acronyms   **233**

About the Authors   **237**

Index   **239**

Foreword

Analog integrated circuits are the interface between electronic systems and the outside world. Generally, they are used as interface for signal sensing, in communication receivers and transmitters, and in output drivers. Since the information resides in the amplitude of the analog signals, these circuits are very vulnerable to all interferences that affect the signal, ranging from noise to crosstalk or any other kind of internal or external interference. At the same time, cost reduction is a driving factor for most applications today, especially in the consumer market. An important means to reduce cost is through the integration of functions. This is the basic motivation for the on-chip integration of more and more complex systems, including both analog and digital circuit functionality, ultimately resulting in fully integrated systems on a chip (SoC). Examples are embedded data converters, single-chip transceivers, hard-disk or DVD readout chips, and so forth. In all of these systems, the digital circuits (including processors and memories) form the bulk of the chip, while the analog interface circuits occupy a relatively small fraction of the area. Yet it is the analog circuits that are vulnerable to interference. One particularly important source of interference in mixed-signal systems that integrate both analog and digital functions on the same die is substrate noise. The switching nature of the digital circuits creates current spikes and resulting voltage variations on supply and substrate. In particular, the substrate voltage fluctuations propagate towards the analog circuits embedded on the same die and deteriorate, and sometimes destroy, the performance of these circuits. This book addresses this important problem of substrate noise coupling in analog and RF systems. It investigates in detail the mechanisms of creation and propagation of substrate noise and analyses its impact on analog circuits. Emphasis is on the modeling and simulation of the effects, as needed by circuit designers during their designs, as well as on design measures to alleviate the problems. Chapter 2 focuses on the methods to simulate switching noise propagation through the substrate, comparing both finite difference and finite element methods. Chapter 3 discusses different types of passive isolation structures and their effectiveness, including guard rings and triple-well shielding. Chapter 4 describes the different noise coupling mechanisms into an active device and verifies this experimentally. Chapter 5 presents measurements to identify the dominant substrate noise couplings in an LC-tank VCO, including the coupling between a power amplifier and the VCO. Chapter 6 outlines simulation methodologies to predict the impact of substrate noise on analog/RF systems. This is then used to analyze different techniques for reducing the noise coupling. Finally, Chapter 7 presents examples of simulated and experimentally verified substrate noise impact on actual circuits such as a millimeter-wave VCO and a wideband receiver. This book therefore provides a guide to all practicing mixed-signal designers who want to obtain insight into the complicated phenomenon of substrate noise coupling into their analog/RF circuits. They will learn how to simulate these things and how to analyze the effectiveness of the design measures that they should take. In this way, they will be able to design better and more robust mixed-signal systems, which makes this book an indispensible guide for all true mixed-signal designers.

Georges Gielen
Full Professor
Katholieke Universiteit Leuven
Leuven, Belgium
March 2010

Preface

Semiconductor companies often struggle with signal integrity problems. During the measurements of their systems on a chip (SoC), spurs pop up that were not expected. Due to those spurs they might not meet the mask requirements and thus an expensive redesign is needed. It is not a trivial task for the analog/RF designer to reveal the origin of those spurs because those spurs are the result of (complex) parasitic coupling effects. One of the most difficult and challenging coupling paths to characterize is the coupling through the common substrate. In lightly doped substrates, the coupling through the substrate is complex and strongly depends on the layout of the SoC.

Our research group at IMEC has more than 10 years of expertise in the analysis and prediction of substrate noise coupling. The results and the latest developments in the field of substrate noise analysis are bundled in this work. This book tackles the problem of substrate noise propagation and the analysis and prediction of the impact of substrate noise in sensitive analog/RF building blocks and systems.

Due to the high resistive nature of the lightly doped substrate, the propagation of substrate noise is complex and not easily predictable with some theoretic expressions. Different mesh-based solutions are proposed, and what is required to model the propagation of substrate noise accurately is shown through a number of practical examples. The most straightforward way to attenuate the propagation of substrate noise is to use guard rings. Mainstream guard rings are discussed and compared. This results in some practical guidelines in how to correctly design such a guard ring.

This book focuses on the analysis of the impact of substrate noise in analog/RF circuits. The analog/RF circuit has many substrate sensing nodes, which each have a sensitivity to substrate noise perturbation that depends not only on the circuit topology but also on the layout of the analog/RF circuit. This makes the analysis of substrate noise very challenging. When the designer notices substrate noise problems during the measurement campaign of his IC, different measurement techniques should be combined in order to isolate the different coupling paths such that their importance can be quantified. Also, different experiments should be set up to show the influence of a certain circuit parameter on the impact of substrate noise. This is not an easy task, but it is the only way to make sure that the correct coupling mechanism is selected. The analog/RF circuit can efficiently be shielded against substrate noise only when the correct coupling mechanism is selected. Moreover the type of measurement experiment that should be set up depends strongly on the analog/RF circuit itself and the external connections that are available. Through this book it is shown that the combination of different measurement techniques allows the extraction of a large amount of information about the substrate noise coupling mechanisms.

In this book, emphasis is also put on the prediction of substrate noise in analog/RF circuits and systems. The importance of the coupling mechanisms does not only depend on the topology of the circuit itself, but also on the layout. Including all of the small layout details of an analog/RF system is not possible. Therefore, only the layout details that determines the dominant coupling mechanism need to be included in the simulation methodology in order to predict the impact of substrate noise correctly. The designer is shown through a number of practical silicon examples how to select the layout details that contribute to the impact of substrate noise. When the coupling mechanisms are identified, the designer can effectively shield his or her analog/RF circuit and make sure that the SoC does not suffer from substrate noise problems.

Acknowledgments

This book bundles more than 10 years of research in the field of substrate noise. Good research can only originate in a fruitful working environment. Therefore we express are sincere gratitude to all those who gave their contribution to make this book possible both at IMEC and the Vrije Universiteit Brussel.

We especially thank all the members of the Wireless Group of IMEC who contributed in one way or another to this book. We thank Bob Verbruggen, Karen Scheir, Jonathan Borremans, Charlotte Soens, Kuba Raczkowsky, Michael Libois, Bertrand Parvais, Luc Pauwels, Bjorn Debaillie, Hans Suys, Danny Frederickx, Cathleen De Tandt, Joris Vandriessche, Mark Ingels, Kameswaran Vengattaramane, Pieter Crombez, Mandal Gunjan, Vojkan Vidojkovic, Vito Giannini, Giovanni Mangraviti, Julien Ryckaert, Lynn Bos, Arnd Geis, Boris Come, John Compiet, Tomohiro Sana, Takaya Yamamoto, Piet Wambacq, Jan Craninckx, Filip Louagie, Pol Marchal, Rudy Lauwereins, Dimitri Linten, Steven Thijs, Mirko Scholz, Xavier Rottenberg, Daniella Van Ravesteijn, Myriam Janowsky, Kristof Vaesen, Wouter Ruythooren, Liesbet Van Der Perre, Stephane Donnay, Fre Vanaverbeke, and Nele Van Hoovels.

We also acknowledge the members of the VUB for the constant trust, confidence, and support. We thank Wendy Van Moer, Johan Schoukens, Rik Pintelon, Kurt Barbé, John Lataire, Lieve Lauwers, Zhi Li, Griet Monteyne, Laurent Vanbeylen, Mattijs Van de Walle, Anne Van Mulders, Liesbeth Gommé, Koen Vandermot, Ludwig De Locht, Maite Bauwens, Veerle Beelaerts, Ann Pintelon, Mohamed El-Barkouky, Sven Reyniers, and Wim Foubert.

Furthermore we also thank the Europractice IC service for fabricating the circuits and the Institution for the Promotion and Innovation through Science and Technology in Flanders (IWT-Vlaanderen) for their financial support.

Chapter 1

Introduction

1.1 INTRODUCTION AND MOTIVATION

The consumer electronic market is mainly driven by cost reduction. This cost reduction is facilitated by the continuous miniaturization of the transistor dimensions. More and more functionality can be integrated on the same die. This leads to the proliferation of single-chip radio implementations which are also called systems on a chip (SoC). Such an SoC is for example present in a PDA, which includes wireless Internet access, mobile phone capabilities, GPS, cameras, touchscreen technology, and various sensors. In such devices, analog/RF and digital functionality are integrated on the same die. Although most of the functionality is implemented by the digital or digital signal processing (DSP) circuitry, analog/RF circuits are still needed at the interface between the digital system and the outside (analog) world. Hence they are both integrated on the same die for cost and performance reasons. Modern SoC designs are therefore more and more mixed-signal. This will be more prevalent if we move toward intelligent homes, health care monitoring, the automotive industry, and so forth. For example such advanced SoC is a wireless sensor network with which body temperature, heart rate, and blood pressure can be monitored anytime for high-risk individuals. This device is then wirelessly connected to the health care provider, who can trace the patient via satellite navigation. This example shows that the integration of the analog/RF circuitry, different sensors, and digital circuitry offers many benefits and possibilities. Unfortunately, the integration of the analog and digital circuits also creates technical problems, especially in deep submicron CMOS technologies. Since the analog circuits exploit the low-level physics of the fabrication process, they are costly and difficult to design, and

1

are also vulnerable to any kind of noise coupling. The higher level of integration, which includes adding more and more digital functionality that is clocked at an ever increasing frequency, makes the mixed-signal signal integrity even more challenging. One of the most important problems is the substrate coupling, caused by the fast switching of the digital circuitry via the common substrate. The continuously ongoing increase in speed and complexity of the digital circuitry on mixed-signal integrated systems also mean an increase of the amount of digital switching noise generated by this circuitry.

On a lightly doped substrate, the most important substrate noise generation mechanism is the coupling of power supply noise into the substrate [1]. When large digital circuits are switching simultaneously, high current peaks are drawn from the power supply network, which results in ringing on the supply lines. This ringing, commonly called switching noise, is injected into the substrate in a resistive way through the substrate contacts. Those substrate contacts connect the ground interconnect of the digital circuits to the substrate. The switching noise then propagates through the common substrate, and hence it is called substrate noise. When substrate noise couples into an analog/RF circuit residing on the same die, it can severely affect the circuits performance. The performance and the precision levels required from the analog/RF circuit increase as dictated by today's applications such as WLAN, UMTS, and so forth. This goes together with an increase of the sensitivity or the susceptibility of the analog circuits to digital substrate noise. It is therefore important to be able to predict the impact of substrate noise on the analog circuit performance at the design stage of the integrated system, before the chip is taped out for fabrication. There are three aspects to such a substrate noise analysis and simulation methodologies (see Figure 1.1):

- Generation of substrate noise;

- Propagation of substrate noise;

- Impact of substrate noise.

The first is the modeling of the digital switching noise injected into the substrate. The amount of switching noise injected into the substrate depends on the switching activity (the amount of switching) of the digital gates, the steepness of the transition time, and the impedance of the power supply network and the package. The higher the value of the inductance of the power supply network and the package, the worse the ringing on the power supply lines and the more switching noise is injected into the substrate. The higher the value of the decoupling capacitors, the better the ringing is damped [2]. In this book, the digital circuitry is replaced by a

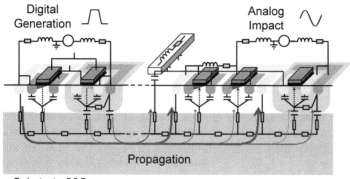

Figure 1.1 The substrate noise analysis can be divided into generation, propagation, and impact.

substrate contact. A substrate contact is a resistive connection to the substrate and is connected via a bond pad to the external world. Any signal can be applied to the substrate through such a substrate contact in a very controlled way.

The second aspect is the propagation of substrate noise through the lightly doped substrate from the aggressor (the substrate contact, the digital circuitry) to the reception point (the victim analog circuitry embedded in the same substrate). This requires a model of the substrate. A typical lightly doped substrate usually consists of a 300 μm high resistive substrate (20 Ωcm) with a 1.5 μm conductive p-well layer (800 S/m) on top. In such a lightly doped substrate, the propagation of substrate noise is strongly influenced (sometimes determined) by the adjacent layout details such as P- and N-doped regions.

Finally, the third aspect of the substrate noise analysis is the modeling of the impact of substrate noise on the analog circuitry. The analog circuitry is not a single noise reception point, it has many noise sensing nodes that all have a different sensitivity to the noise. Such an analysis is quite complex and time consuming for large analog circuits such as entire front ends. Hence, some simplifications might be needed in order to make this analysis tractable.

In addition to the analysis methodology, design guidelines and techniques to reduce the substrate noise problem are developed. This can be done by either decreasing the transmission or increasing the immunity of the analog circuits. Guidelines are developed to increase or alter the transmission between the aggressor and the victim by using guard rings. The immunity of the analog circuitry to substrate noise can be effectively increased by shielding the dominant substrate noise entry points. To this end, several layout and circuit techniques are proposed.

1.2   BOOK OVERVIEW

The purpose of this book is to give the analog designer a good understanding about how substrate noise propagates through the substrate and how it couples into the analog circuitry. A deep understanding is provided through numerous examples. Further, guidelines are provided to the designer to tackle substrate coupling problems. Some of these guidelines should be taken into account already at the architectural level.

To guarantee first silicon pass, it is mandatory to accurately predict the impact of substrate noise in the sensitive analog circuitry. To this end, a methodology is presented that combines both the strengths of the emerging 3D field solvers and the circuit simulators. A large number of examples are modeled with this methodology, which are all validated with measurements.

This book is structured as follows:

- Chapter 2 gives a deep understanding of how substrate noise propagates through the substrate and how small layout details like n-wells influence this propagation. Insight into the substrate noise propagation will determine what is required to bring simulations in agreement with measurements. Consequently, different techniques are proposed that accurately model the substrate.

- Chapter 3 proposes different isolation structures, also known as guard rings, to increase the isolation between the aggressor (digital circuit) and the victim (analog circuit). The substrate noise isolation provided by the different guard rings are compared. A properly designed guard ring can improve the substrate noise isolation by more than 40 dB. Guidelines are formulated to design efficient guard rings.

- Chapter 4 proposes a simulation methodology to predict the impact of substrate noise in active devices such as transistors. The different substrate noise coupling mechanisms in a transistor are revealed and compared. It is shown that the impedance of the adjacent interconnects determines the importance of the coupling mechanisms.

- Chapter 5 teaches the designer different measurement techniques that can be used to reveal and quantify the dominant substrate coupling mechanisms. In this way, the different coupling mechanisms between a power amplifier (PA) and a voltage controlled oscillator (VCO) are revealed and quantified. This chapter also introduces the concept of sensitivity functions. These sensitivity functions determine how circuit properties such as gain, oscillation frequency, and so forth vary when a perturbation is applied on a reception node of the circuit.

- Chapter 6 proposes different simulation methodologies that are able to predict the impact of substrate noise in analog/RF circuits with an accuracy that is better than 3 dB. Further, it is shown that the impact of substrate noise mostly results in ground bounce. Different circuit and layout techniques are proposed to increase the immunity of analog/RF circuits to substrate noise. Finally, simulations show that 3D stacking offers an opportunity to reduce the substrate noise coupling.

- Chapter 7 refines the methodology such that it can handle large analog/RF circuits and even analog/RF systems. It is shown that the refined methodology is able to accurately predict the impact of substrate noise in a wideband receiver. The performed simulations are validated with measurements on a silicon prototype. The methodology is able to predict the impact of substrate noise in less than 2 hours with an accuracy that is better than 2 dB [3–5].

References

[1] M. Badaroglu, M. van Heijningen, V. Gravot, S. Donnay, H. De Man, G. Gielen, M. Engels, and I. Bolsens, "High-level simulation of substrate noise generation from large digital circuits with multiple supplies," *Proc. Design, Automation and Test in Europe Conference and Exhibition 2001*, March 13–16, 2001, pp. 326–330.

[2] M. Badaroglu, G. Van der Plas, P. Wambacq, S. Donnay, G. Gielen, and H. De Man, "SWAN: high-level simulation methodology for digital substrate noise generation," *Symposium on VLSI Circuits*, Vol. 14, No. 1, Jan. 2006, pp. 23–33.

[3] S. Bronckers, K. Scheir, G. Van der Plas, and Y. Rolain, "The impact of substrate noise on a 48-53 GHz mm-wave LC-VCO," *Proc. IEEE Topical Meeting on Silicon Monolithic Integrated Circuits in RF Systems SiRF '09*, 2009, pp. 1–4.

[4] S. Bronckers, C. Soens, G. Van der Plas, G. Vandersteen, and Y. Rolain, "Simulation methodology and experimental verification for the analysis of substrate noise on LC-VCO's," *Proc. Design, Automation & Test in Europe Conference & Exhibition DATE '07*, 2007, pp. 1–6.

[5] S. Bronckers, G. Vandersteen, C. Soens, G. Van der Plas, and Y. Rolain, "Measurement and modeling of the sensitivity of LC-VCO's to substrate noise perturbations," *Proc. IEEE Instrumentation and Measurement Technology*, 2007, pp. 1–6.

Chapter 2

Substrate Noise Propagation

2.1 INTRODUCTION

Current deep submicron technologies use a lightly doped substrate (20 Ωcm). This type of substrate has better substrate noise isolation properties than the epitype substrate [1–3] (0.1 Ωcm) because the lightly doped substrate has a higher resistivity. However, the substrate noise propagation in a lightly doped substrate is much more complex [4] and depends on the layout [5].

The goal of this chapter is to get a better understanding of the signal propagation in a lightly doped substrate and to determine what is required to model the substrate noise propagation accurately. It will be shown that for very simple geometrical structure simple analytical formulas exist. Those formulas can give insight into how substrate noise currents flow. However, for more complex geometrical structures the designer needs to rely on simulation tools. Different simulation tools exist that describe the propagation of substrate noise [6–12]. Most of these simulation tools implement either the finite difference method (FDM) or the finite element method (FEM) [13] (see Table 2.1). Other techniques [12, 14] are surfacing that address this complex modeling issue.

To determine what is required to model the substrate noise propagation accurately, both methods are demonstrated on a small silicon prototype. FDM is implemented by the simulation tool $SubstrateStorm$ [15] and FEM is implemented by $HFSS$ [16]. The designer is taught how to use the tools correctly. Both tools give insight into how substrate noise propagates through the substrate and how layout details, such as metal traces and P^+/N^+ doped regions, influence the substrate noise propagation. The performance of both tools are compared against each other

Table 2.1

The Simulation Tools Can Be Categorized into FDM and FEM

FDM	FEM
SubstrateStorm [6]	HFSS [9]
SeismIC [7]	MEDICI [10]
SPACE [8]	Sequoia [11]

and against measurements in terms of accuracy, a priori knowledge, and simulation time. Both tools are able to describe the substrate noise propagation accurately, but the HFSS simulation tool that implements FEM is much faster. Therefore, this tool is used in Chapter 3 to characterize the substrate noise propagation in complex isolation structures.

2.2   MODELING THE SUBSTRATE

In order to model the substrate, it usually suffices to solve the Maxwell equations. The current flow of the free carrier can be neglected due to the homogeneity of the semiconductor [17].

The differential form of the Maxwell equations are [18]:

$$\nabla \vec{E} = \frac{\rho_c}{\epsilon} \tag{2.1}$$

$$\nabla \times \vec{E} = -\frac{\partial \vec{B}}{\partial t} \tag{2.2}$$

$$\nabla \vec{B} = 0 \tag{2.3}$$

$$\nabla \times \frac{1}{\mu} \vec{B} = \vec{J} + \epsilon \frac{\partial \vec{E}}{\partial t} \tag{2.4}$$

where $\vec{E}$ is the electric field vector, $\vec{B}$ is the magnetic field vector, ∇ is the Laplacian, ρ_c is the charge density, ϵ is the dielectric constant, $\vec{J}$ is the current density vector, and μ is the permeability of the semiconductor material.

Starting from the Maxwell laws, a qualitative analysis is given to extract the resistance between two substrate contacts. In order to handle more complex geometries, the designer needs to rely on numerical methods. To that end, this section briefly rephrases how FDM and FEM work in order to instruct the reader on how the different tools model the substrate. The tools differ from each other in the way that they solve those Maxwell equations. FDM approximates the differential Maxwell equations by finite difference equations and then solves these equations. FEM approximates the solution of the differential form of the Maxwell equations. An in-depth understanding of FDM and FEM can be found in [19]. In the next section, both methods are implemented by a different tool on the same simple silicon example. This allows us to assess the performance of the different tools.

2.2.1 Analytical Resistance Calculation Between Two Contacts

This section provides analytical formulas to extract the substrate resistance between two contacts. Those analytical formulas give the designer insight into how the current flows into the substrate but is restricted to very simple geometrical structures like two rectangular or circular contacts. The resistance between two rectangular contacts are discussed from the point of view of the substrate noise current flow.

In the case of a one-dimensional current flow, the current can be considered as flowing in a floating well with two contacts at either side, or simply a resistor. In order to calculate the resistance between the two ends of the floating well, Maxwell's equations need to be solved. Since only the resistance is of interest and a quasi-static (infinitely slow, the charges are in equilibrium) solution can be assumed, one needs to solve the first law of Maxwell, also called the Poisson equation:

$$\nabla \vec{E} = \frac{\rho_c}{\epsilon} \tag{2.5}$$

In the case that no charges are present, the Poisson equation can be simplified to:

$$\nabla \vec{E} = 0 \tag{2.6}$$

The corresponding current density is proportional to the electrical field and inversely proportional to the resistivity of the layer (ρ_{layer}). If one assumes that the current density and the electrical field are constant across the floating n-well this becomes:

$$J = \frac{E}{\rho_{layer}} \tag{2.7}$$

The current (I) through the floating well is given by the surface (S) integral of the current density. Consider a rectangular volume with dimensions t_{layer}, w_{layer}, and d (see Figure 2.1). The current I is flowing from left to right.

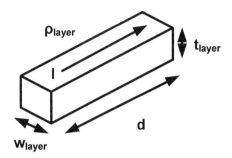

Figure 2.1   Floating well with dimension t_{layer}, w_{layer}, and d.

In this case, the current I is equal to:

$$I = \oint_S J \cdot dS = w_{layer} \cdot t_{layer} \cdot J \qquad (2.8)$$

Substituting (2.8) in (2.7) gives:

$$E = \nabla V = \rho_{layer} \frac{I}{w_{layer} \cdot t_{layer}} \qquad (2.9)$$

Hence:

$$V = \int_0^d \rho_{layer} \frac{I}{w_{layer} \cdot t_{layer}} = \rho_{layer} \cdot \frac{I \cdot d}{w_{resistor} \cdot t_{layer}} \qquad (2.10)$$

Then the resistance value depends on the length of the resistor (d) and its area

.

$$R_{resistor} = \frac{V}{I} = \rho_{layer} \cdot \frac{d}{w_{resistor} \cdot t_{layer}} \qquad (2.11)$$

Equation (2.11) is also called the law of Pouillet. In this case, there exists a linear relationship between the resistance between two contacts and the distance between those contacts. ρ_{layer}/t_{layer} is also called the sheet resistance R_{sheet}. The sheet resistance is typically used to calculate the resistance of rectangular sheets of material in terms of number of squares.

In the case of two-dimensional current flow, the resistance depends on the distance over size ratio. An example of this is the resistance of a thin layer (in this case a well) [20]:

$$R_{resistor} = \frac{\rho_{layer}}{\pi \cdot t_{layer}} \cdot (ln(\frac{d}{r}) + 0.25) \tag{2.12}$$

where r is the radius of the substrate contact.

This formula is a good approximation for a layered situation. It assumes that the top layer is much more conductive than the bottom layers and that the layer extends until infinity in an homogeneous way. Current submicron technologies are using a lightly doped substrate. A typical lightly doped substrate consists of a 300 μm high resistive substrate (20 Ωcm) with a 1.5 μm p-well ($\pm$ 0.2 Ωcm) on top. Thus the p-well layer is two orders of magnitude more conductive than the intrinsic material. If the substrate contacts are closely spaced, this formula provides a good first order estimation of the resistance between both contacts.

In the case of three-dimensional current flow, the resistance depends on the distance over size ratio, but only for small distances. An example of this is the infinite slab resistance [21, 22]:

$$R_{resistor} = \frac{\rho_{layer}}{2r} \cdot (1 - \frac{2}{\pi} arcsin \frac{r}{d+r}) \tag{2.13}$$

A Taylor expansion of (2.13) shows that to first order the value of the resistance is proportional to $1/d^2$.

$$R_{resistor} = \frac{\rho_{layer}}{2r} \cdot (1 - \frac{1}{3\pi} \cdot (\frac{r}{d+r})^2) \tag{2.14}$$

The underlying explanation of this is that the current density is either:

- Constant for the case of one-dimensional flow;

- $1/d$ for the case of two-dimensional flow;

- $1/d^2$ for the case of three-dimensional flow.

This once again is caused by the fact that the area available for the current to flow through is constant for one-dimensional flow, proportional to the distance (circumference of the circle for a two-dimensional flow), or proportional to the distance^2 (surface area of a sphere for a three-dimensional flow). The potential, which is the integral of the current density, changes linearly (one-dimensional), logarithmically (two-dimensional), or with arcsin (three-dimensional) and is directly related to the resistance given a constant current.

It is now possible to qualitatively analyze how the current flow is in some situations. Let's consider the case of a single contact on top of a layered substrate,

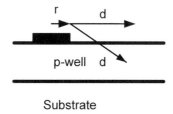

Substrate

Figure 2.2   Lightly doped substrate, contact size, and distances.

shown in Figure 2.2. The radius of the contact is given by r and the distance to a position on the well or in the material is given by d. Both the substrate and the well have a resistivity, given by ρ_{well} and ρ_{sub}, respectively. Remember that in the case of a lightly doped substrate the ρ_{well} is two orders of magnitude lower (thus more conductive) than ρ_{sub}.

If current is injected into the contact, the current can flow in two ways. It can either spread in a two dimensional way (in the well) or go straight to the substrate (assume that this is possible for the qualitative reasoning sake) and spread in a three-dimensional way. It is reasonably easy to determine which path the current will take. If we take in this particular case the incremental resistance dR as a measure for the attractiveness of the path (i.e., a low incremental resistance means the resistance will increase more slowly, or the current will more easily take that path), then we have to compare the incremental resistance through the well to the incremental resistance through the substrate:

$$dR_{well} = \rho_{well} \cdot \frac{dx}{2\pi dt_{well}} \qquad (2.15)$$

$$dR_{sub} = \rho_{sub} \cdot \frac{dx}{2\pi d^2} \qquad (2.16)$$

The distance d at which the last two quantities are equal is:

$$d_{2D3D} = t_{well} \cdot \frac{\rho_{sub}}{\rho_{well}} \qquad (2.17)$$

d_{2D3D} is called the two-dimensional, three-dimensional corner. The value of the corner distance is now calculated in the case of a lightly doped substrate with a substrate and p-well resistivity of, respectively, 20 Ωcm and 0.2 Ωcm. Next, the current flow in a lightly doped substrate is discussed and it is shown that the applicability of the above-mentioned formulas is limited. The corner distance in the case of a lightly doped substate is:

$$d_{2D3D} = 1.5 \ \mu m \cdot \frac{20 \ \Omega cm}{0.2 \ \Omega cm} = 150 \ \mu m \qquad (2.18)$$

Assume first that the radius r of the contact is larger than this two-dimensional, three-dimensional corner of 150 μm. In this case the current has a better conducting path through the substrate than through the well path.

Assume the radius r is smaller than this corner distance. Then at first the current will spread in the well, but from the corner distance the amount of the current flowing in the well will decrease, since the path through the substrate has a lower resistance.

At this point it is probably appropriate to warn that this is only a qualitative understanding. The two-dimensional current flowing through the well has to spread to a three-dimensional current flow in the substrate. This will not happen immediately beyond the corner distance. It cannot be determined with this reasoning how fast this will happen.

This reasoning does suggest that for large structures the current will (except for small distances compared to the size of the contact region) primarily flow through the substrate to other contacts on the substrate. For small contacts, the current will flow through the well layer at first, but when the distance is increased, the current will increasingly flow through the substrate. In practice, the corner value of 150 μm is a realistic value for the separation between the digital and analog circuitry. Hence, only numerical tools can determine the amount of current that flows through the p-well and the substrate. Furthermore, those formulas are restricted to only two substrate contacts and very simple geometrical structures (no adjacent wells are present). The next two sections explain how the substrate can be modeled with the finite difference and the finite element method.

2.2.2 Finite Difference Method

The basic principle [23, 24] behind this modeling is explained in this section. The substrate, without the diffusion/active areas, is considered to consist of layers of uniformly doped semiconductor material of varying doping profiles. In these layers, a simplified form of Maxwell's equations is formulated:

$$\frac{\partial}{\partial t}(\nabla \cdot \epsilon \cdot E) + \nabla \cdot \frac{1}{\rho} \cdot E = 0 \qquad (2.19)$$

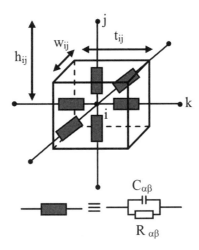

Figure 2.3   Cube around node i in the box integration technique.

Hereby the influence of the magnetic fields are ignored. There are several ways to solve (2.19). In the FDM technique, the substrate is represented as a collection of square cubes as shown in Figure 2.3. The electrical field normal to a contact plane of two adjacent cubes (i,j) with distance h_{ij} is given by:

$$E_{ij} = \frac{V_i - V_j}{h_{ij}} \tag{2.20}$$

with V_i and V_j the potential on the nodes i and j. h_{ij} is the distance between nodes i and j. Using Gauss's law under the assumptions of a uniform doping concentration in the box integration method, it is possible to rewrite (2.19) as:

$$\sum_j [\frac{(V_i - V_j)}{R_{ij}} + C_{ij}(\frac{\partial V_i}{\partial t} - \frac{\partial V_j}{\partial t})] = 0 \tag{2.21}$$

where

$$R_{ij} = \rho \cdot \frac{h_{ij}}{w_{ij} t_{ij}} \tag{2.22}$$

$$C_{ij} = \epsilon \cdot \frac{w_{ij} t_{ij}}{h_{ij}} \tag{2.23}$$

are used in the lumped model in Figure 2.3. The number of lumped elements depends on the electrical field intensity. Although the electric field varies nonlinearly as a function of distance, the box integration method approximates this variation as a piecewise constant function. In regions where the gradient of the electric field is high, it is necessary to use fine grids to accurately approximate the nonlinearity of the electric field. Hence, more lumped elements are needed to correctly describe the substrate noise propagation. Elsewhere, coarse grids can be used to reduce the overall number of grids. However, since the field intensity cannot be determined before discretization, the density of the grid needed is not known a priori.

The generated dense RC model can be approximated for a certain frequency range by a smaller network that exhibits similar electrical properties by using a model order reduction technique [25].

At sufficiently low frequencies, the RC model can be further simplified without losing accuracy. First of all, the substrate can be approximated at sufficiently low frequencies by a resistive mesh only. In order to determine at which frequencies the capacitive behavior of the substrate should be taken into account, the relaxation time of the bulk substrate outside of the active areas and the well diffusions is calculated. This relaxation time is given by $\tau_{SUB} = \rho_{SUB}\epsilon_{SUB}$ and is of the order of 10^{-11} sec (for $\rho_{SUB} = 15$ Ωcm and $\epsilon_{r,SUB} = 11.9$, $\epsilon_0 = 8.8510^{-12}$ F/m). Hence, it is reasonable to neglect intrinsic substrate capacitances for operating speeds of up to a few gigahertz. From (2.23) the following relation can be derived for the capacitances and resistances in the bulk substrate:

$$Rij \cdot Cij = \rho_{SUB}\epsilon_{SUB} = \tau_{SUB} \qquad (2.24)$$

The cutoff frequency at which the impedance associated to the capacitance C_{ij} becomes comparable to the resistance R_{ij} is [26]:

$$f_{c,SUB} = \frac{1}{2\pi\tau_{SUB}} = \frac{1}{2\pi \cdot \epsilon_{SUB} \cdot \rho_{SUB}} = \frac{1}{2\pi C_{ij}R_{ij}} \qquad (2.25)$$

Table 2.2

Cutoff Frequency for Different Substrate Doping Levels in a Typical Lightly Doped Substrate [27]

Doping level (cm^{-3})	Resistivity ρ_{SUB} (Ωcm)	f_c^{SUB} (GHz)
10^{14}	40	3.75
10^{15}	5	30
10^{16}	0.5	>1,000

Table 2.3

Skin Depth Versus Frequency for a Typical Lightly Doped Substrate [27]

Frequency (GHz)	T_{skin} in a lightly doped substrate, 10 Ωcm resistivity (μm)
3	2,906
7.5	1,838
15	1,300

As an example for the 20 Ωcm substrate used in this work $f_{c,SUB}$ equals 7.5 GHz. Other examples are given in Table 2.2. For noise frequencies below 7.5 GHz, a resistive mesh suffices.

Second, the skin-effect can also be neglected at sufficiently low frequencies. The skin-effect is the phenomenon where at higher frequencies the signals do not penetrate in the substrate any more but stay at the surface of the substrate within the skin depth, labeled T_{skin}. The field penetration depth or skin depth in the substrate is given by the approximate expression [27]:

$$T_{skin} = \sqrt{\frac{\rho_{SUB}}{\pi \mu_{SUB} f}} \qquad (2.26)$$

Here μ_{SUB} is the substrate permeability [its value is $4\pi 10^{-7}(H/m)$], ρ_{SUB} is the substrate resistivity, and f is the frequency of the signal propagating in the substrate. For epi-type substrates, the skin depth is of the order of the wafer thickness or even lower. In general, since the substrate currents in epi-type substrates flow concentrated just under the epitaxial layer, the skin effect has an influence at frequencies higher than:

$$f = \frac{\rho_{SUB}}{\pi \mu_{SUB} T_{skin}^2} \qquad (2.27)$$

The wafer thickness used in all our examples is 11 mil (≈ 300 μm). As can be seen in Table 2.3, the skin effect will not be a problem in the lightly doped substrates at the noise frequencies studied in this work.

Once the substrate itself is modeled, the capacitance from the depletion regions of well diffusions are modeled. Those are usually modeled outside the mesh using lumped elements [15].

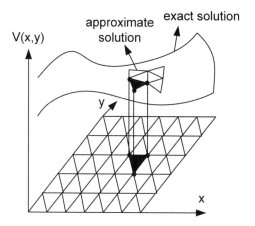

Figure 2.4 The FEM seeks a solution that approximates the exact solution best.

2.2.3 Finite Element Method

The finite element method can be used to generate an electromagnetic field solution for the substrate [19]. The solution is a continuous piecewise function that approximates the exact solution best [19]. The solution can be either a potential or electromagnetic field values. In order to instruct the reader how this approximated solution is found, assume a two-dimensional problem (see Figure 2.4) where the solution, which is a priori not known, is given by:

$$V = V(x, y) \qquad (2.28)$$

where V is a continuous potential function. Its value depends on the spatial coordinates x and y. The two-dimensional space is meshed with triangles, which are called elements (see Figure 2.4). The combination of triangles is referred as the finite element mesh. Each triangle has three corners and each corner has a corresponding voltage V_i. The corresponding voltages of the three corners of a triangle form a plane, which approximates the exact solution in that region. The different planes, formed by the voltages of the different triangles, are a good approximation for the exact solution. The goal is to calculate the values of V_i that approximate the exact solution best. Therefore, the voltage function in each triangle is written as a weighted sum of the voltages at the corners of each triangle. The voltage function in region k which corresponds to triangle k can be written as:

$$V_k(x, y) = \sum_{i=1}^{3} V_{i,k}\alpha_{i,k}(x, y) \tag{2.29}$$

The voltage function $V_k(x, y)$ corresponds to the voltages $V_{i,k}$ at the corners of triangle k weighted with a set of basis functions, referred as $\alpha_{i,k}$. Those basis functions define the solution inside the triangle. The constraints on the function $V(x, y)$ are that it needs to obey to the Maxwell's laws and the boundary conditions. The boundary conditions correspond to a voltage on a corner of a triangle with a user-given value.

The laws of Maxwell are incorporated in the basis functions. For simplicity, we demonstrate how FEM enforces that the voltage function $V(x, y)$ obeys the first law of Maxwell in a region that has a constant charge density (ρ):

$$\nabla^2 V(x, y) = \frac{\rho}{\epsilon} \tag{2.30}$$

Substituting (2.29) in (2.30) gives for triangle k:

$$V_k(x, y)\nabla^2 \alpha_{i,k}(x, y) = \frac{\rho}{\epsilon} \tag{2.31}$$

Thus, the potentials at the corners of triangle k are weighted in such a way that they obey Maxwell's law. In this way, the voltages at the corners of all the other triangles can be solved since the basis functions are known a priori.

The mesh size is determined by the error between the shared corner of two adjacent triangles. If the error is larger than specified in the error criteria, the mesh needs to be refined. When the error criteria is met, this piecewise triangulated function is a good approximation to the exact solution.

In a three-dimensional space, like in HFSS [16], the elements are represented by tetrahedra. A tetraheder is a four-sided pyramid as shown in Figure 2.5. This collection of tetrahedra is also referred to as the finite element mesh. The field quantities are solved at the nodes of the tetrahedra. In HFSS, the interface between the EM simulator and the external world is formed by ports. The S-parameters are solved at the boundaries of the ports.

The size of the mesh is a trade-off between speed and accuracy. The mesh is refined until the maximum ΔS criterion is satisfied. The maximum ΔS is the maximum difference of S-matrix magnitudes at the ports between two consecutive passes. If the difference in magnitudes of the S matrices changes by an amount less than the maximum ΔS value from one pass to the next, the maximum ΔS criterion is satisfied.

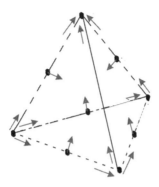

Figure 2.5 The field quantities are solved at the nodes of the tetraheder.

FEM can be used to solve the substrate. The substrate is also assumed to be uniformly doped. In this case both the electrical and magnetic fields are solved. Possible skin effects in the substrate are also automatically taken into account although the previous section pointed out that this is not necessary for low frequency substrate noise.

2.3 THE SUBSTRATE MODELED WITH FDM

By the means of a simple test structure, it is shown that the substrate can be accurately modeled by the simulation tool SubstrateStorm, which implements FDM. The substrate noise propagation on this test structure is interpreted. Simulations show that layout details, such as n-well regions and metal traces, influence the substrate noise propagation. N-wells are attracting the substrate currents. Metal traces pick up substrate currents and inject them again in the substrate at another location.

2.3.1 Experimental Description

The goal of this experiment is to gain insight in the substrate noise propagation in an lightly doped substrate and to determine to which extent layout details influence the substrate noise propagation. Therefore, a simple test structure is designed that contains most of the layout features such as n-wells and metal traces that are present in any analog circuit. The test structure and its most important features are shown in Figure 2.6. It contains two resistive contacts to the substrate with a size of 40

Table 2.4

Dimensions of the Test Structure Shown in Figure 2.6 [29]

d_1	d_2	d_3	d_4	d_5
250 μm	500 μm	150 μm	125 μm	125 μm
w_1	w_2	w_3	$w_{contact}$	$h_{contact}$
100 μm	80 μm	150 μm	40 μm	20 μm

$\times$ 20 μm^2 (black filled rectangles), indicated with S$_1$ and S$_2$. In this experiment the substrate noise propagation between those two substrate contacts is modeled and then measured to assess the accuracy of the obtained model. Also shown in Figure 2.6 are the lowest level metal wires present in the vicinity of the contacts. These wires pick up signal from the substrate through capacitive coupling as will be discussed in Section 2.3.2. Metal wires L, R, and Z are grounded during the S-parameter measurements. Metal wires A, B, and C are left floating. Also shown are two n-wells (indicated with X and Y) at the right edge of the die. The n-wells X and Y are connected to ground via large decoupling capacitors. These peripheral structures are part of a substrate noise sensor [28] (on top of wire R) that is not further used in this experiment. The dimensions of the test structure are summarized in Table 2.4

The experiment is performed in a 1P5M 0.25 μm twin-well CMOS process, with a lightly doped substrate (approximately 18 Ωcm). The sheet resistance of the n-well is approximately 400 $\Omega/\square$ and the sheet resistance of the p-well is approximately 800 $\Omega/\square$. Both the n-well and the p-well are 1.2 μm thick. The properties of the substrate and the different wells are included in the *doping profile* information. SubstrateStorm needs this doping profile information in order to predict the propagation of substrate noise. For simulating the 45 frequency points of Figure 2.8, a simulation time of 144 hours on an HP-UX L2000/4 (4 PA-8600 at 540 MHz) server is required. The predicted substrate noise propagation is compared in the next section against measurements.

2.3.2 Analysis of the Substrate Noise Propagation

The contact to contact resistance is measured with a standard multimeter. The forward propagation (S$_{21}$), which reflects the substrate noise propagation, is measured with a network analyzer on a Cascade probe station, where an acrylic glass plate is inserted between the die and chuck for isolation as in [30].

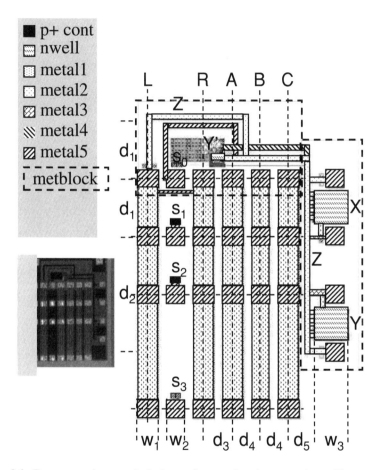

Figure 2.6 Test structure layout: only the layers closest to the substrate are drawn. The propagation between P$^+$ contacts S_1 and S_2 is investigated. The white on gray letters identify metal wires and wells. Metal wires L, R, A, B, and C are split in 3 sections (left on the layout: S_0S_1, S_1S_2, S_2S_3) [29].

2.3.2.1   Contact-to-Contact Resistance

The measured DC resistance between contact S_1 and S_2 is 676Ω. The value obtained by extraction with SubstrateStorm [15] is 613Ω, or a deviation of approximately 8%. A possible explanation for this discrepancy is that the technology parameters provided are nominal values that do not account for process variability. A deviation of p-well resistivity of 10% would for instance explain the discrepancy.

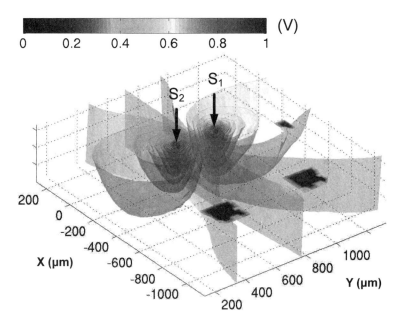

Figure 2.7   Equipotential surfaces at DC in the substrate when 1V is applied to contact S_1 by a 50Ω terminated DC source. Contact S_2 is terminated with a 50Ω resistance. (See color section.)

To get a better understanding of the propagation at DC, a 3D simulation was performed with SubstrateStorm [15] and Spectre [31]. In Figure 2.7 the result of this simulation is shown. A 50Ω terminated DC source of 1V is applied to contact S_1 and a 50Ω load is attached to sink contact S_2. Figure 2.7 shows the equipotential surfaces. The surfaces are spaced linearly on a voltage scale. Close to the contacts S_1 and S_2, the equipotential surfaces are spaced close to each other, or the voltage drop in that region is very high. This signifies a high current density. The surfaces are elliptic at the top of the substrate (p-well). This is consistent with a two dimensional current flow in a thin sheet of conducting material (i.e., the p-well). Note that the p-well at the top of the substrate is two orders of magnitude more conductive than the intrinsic substrate. The equipotential surfaces extend down underneath the contacts. This indicates that the current is penetrating into the lightly doped, highly resistive, substrate layer. Further away the equipotential surfaces are fairly close to vertical. At this point the current flow is horizontal in both p-well and the lightly doped substrate layer.

The above analysis suggests that the current flow in lightly doped substrates between medium size contacts is mostly two dimensional. Only a small portion of the current flows vertically between p-well and the intrinsic substrate layer. This analysis confirms the results in [32] for lightly doped substrates.

Since most of the current flow takes place in a single layer (the p-well layer), the resistance between the two ports can now be approximated by the 2D spreading resistance in a thin conducting layer:

$$R = \frac{\rho_{pwell}}{\pi \cdot t_{pwell}}(0.25 + ln(\frac{d+r}{r})) \tag{2.32}$$

where r is the radius of the contact, d is the distance between the contacts, ρ_{pwell} is the p-well resistivity, and t_{pwell} is the p-well thickness. The formula has been derived for a small thickness ($t \rightarrow 0$) and infinitely extending thin layers. When the rectangular contact is converted into an equivalent circular one by (equal perimeter, because of 2D flow):

$$r = \frac{w_{contact} + h_{contact}}{\pi} \tag{2.33}$$

a resistance value of 770Ω is obtained. This is a reasonable approximation, given that the current flowing in the lightly doped substrate layer has been neglected.

2.3.2.2   S-Parameter Measurement (Forward Propagation)

S-parameter measurements have been performed on the test structure. The forward propagation is shown in Figure 2.8.

At low frequencies (< 10 MHz) the S_{21} propagation is equivalent with a floating 676Ω resistance between the two contacts. From 30 MHz on, the magnitude of the propagation decreases with a slope of approximately 10 dB/decade. From 1 GHz on, the slope increases to 20 dB/decade. Above 5 GHz the propagation magnitude increases again. The simulated forward propagation is shown in Figure 2.9(a). The

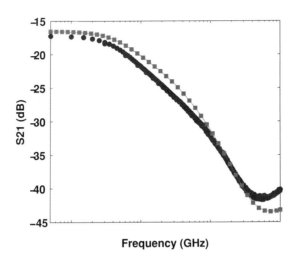

Figure 2.8 Analysis of forward propagation: S_{21} plot of the test structure. The • reflects the measurements and the □ corresponds to simulations.

difference between measured and simulated S_{21} is below 1 dB (approximately 8%) at low frequencies, increases to 2 dB for medium frequencies (100 MHz–3 GHz), and is 3 dB at 10 GHz.

To get a better understanding, a number of 3-D equipotential plots, expressed in dBV, are shown in Figure 2.9. The simulations correspond to a forward propagation measurement: the signal is applied to S_1 and the load is connected to the sink S_2. The first observation is that with increasing frequency, the signal level in the substrate decreases. It suffices to check the scale values on the surface plots.

At 100 MHz, the source and the sink are clearly visible. Also, the n-wells (X and Y) are visible. The equipotential surfaces underneath indicate that current is picked up by the n-wells. In Figure 2.10, the magnitude of the currents picked up by various parts of the test structures are plotted versus frequency. The n-wells are indeed picking up current below 200 MHz. This current does not reach the sink anymore, hence the decrease in S_{21}.

From 100 MHz to 1 GHz, the metal wires pick up more and more current (see Figure 2.10), for a definition of the names see Figure 2.6. On the equipotential plot shown in Figure 2.9(b), at 1 GHz the metal wire sections $L{:}S_2S_3$ and $R{:}S_2S_3$ have silenced the substrate.

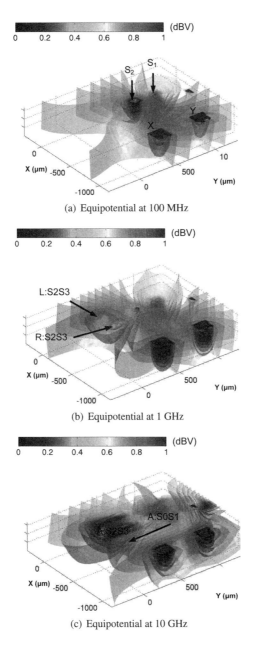

(a) Equipotential at 100 MHz

(b) Equipotential at 1 GHz

(c) Equipotential at 10 GHz

Figure 2.9 3-D equipotential surfaces when 1V is applied to contact S_1 and 0V is applied to contact S_2. Note that the unit is dBV. (See color section)

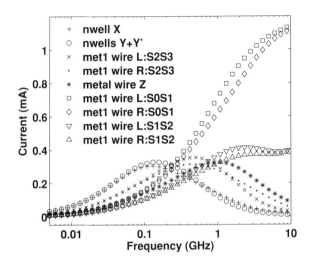

Figure 2.10   Currents picked up by wires and n-wells.

In the 10 GHz plot [Figure 2.9(c)], only the area close to the source is still noisy, and almost all other parts of the structure are silent. The metal wires closest to the source pick up most of the current, shielding the substrate regions further away. The current picked up by the distant wires and n-wells decreases. At approximately 10 GHz, the capacitance between substrate and metal becomes a short, removing any frequency dependency in the (simulated) propagation.

The increase in the propagation at 10 GHz is explained by direct capacitive coupling: the metal 5 bond pad (connected to the source) injects current capacitively into the substrate. A sensitivity analysis revealed that an increase with 20% of the bond pad parasitic capacitance to the substrate increases the propagation by more than 1 dB above 6 GHz.

A last observation is the effect caused by the floating metal wires A, B, and C. On the equipotential plots the substrate underneath these floating wires is noisier than the n-wells (X and Y) and grounded metal wires (L and R). Check for instance position [-750,0] on the 1 GHz plot [see Figure 2.9(a)]. An explanation is found in Figure 2.11. The floating metal 1 wire A picks up current from the substrate near the source (sections $S_0 S_1$ and $S_1 S_2$) and injects this current into the substrate under wire A section $S_2 S_3$. This current is maximal at approximately 1 GHz (0.2 mA), and then decreases because the ground wires L and R shield the source at higher frequencies. Similar observations (albeit with a lower current level) apply for wires B and C in Figure 2.6.

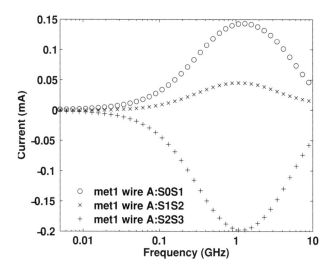

Figure 2.11 Current carried by metal wire A.

2.3.3 Conclusions

In this section, the finite difference method is used to study the substrate noise propagation in a lightly doped substrate. The study of a test structure reveals that the current flow in a lightly doped substrate is mostly two-dimensional (horizontal current spreading). This two-dimensional current flow explains why distant structures (like n-wells and metal traces) pick up currents and influence the propagation. There is also a large interaction between metal and substrate. Currents picked up by metal can be carried to other places and reinjected into the substrate. Taking all these effects into account we have been able to match simulation results with measured data from DC ($< 8\%$ error) up to 10 GHz with a maximum error of 3 dB. The main drawbacks of the simulation tool SubstrateStorm that implements FDM is the very long simulaton time that is required to simulate this simple test structure. In addition, the doping profiles must be provided by the foundries. Those drawbacks are circumvented by the tool HFSS that implements FEM in the next section. The next section models the substrate as a finite element model.

2.4   THE SUBSTRATE AS A FINITE ELEMENT MODEL

This section demonstrates FEM used to study the propagation of substrate noise. By the means of an EM simulator the propagation of substrate noise is studied on the same test structure as the one used to demonstrate FDM. This allows a fair comparison of the performance of the tools that implement both methods. First, the reader is instructed step by step how to use an EM simulator to predict the propagation of substrate noise. Then, by the means of the test structure of the previous section, it is shown that the tool that implements FEM predicts the propagation of substrate noise with the same accuracy as the tool that implements FDM. However, the tool that implements FEM is much faster.

2.4.1   Simulation Methodology

This section briefly explains how an EM simulator can be used to predict the propagation of substrate noise for any passive structure. The simulation methodology consists of three important steps:

1. Preparing the layout for the EM simulator
    The simulation methodology starts from the layout. From a computational point of view, it is neither relevant nor possible to simulate all the small details of the layout. Fortunately, this level of detail is not required to characterize the propagation of substrate noise. Different simplifications are proposed that marginally influence the frequency response but significantly speed up simulations:

    - The different vias that connect the different metal layers are grouped to one single via (see Figure 2.12). The impedance of the interconnect with and without grouped vias may not change drastically. Otherwise either the area of the via or the conductivity of the via's material needs to be changed. A typical example where the vias can safely be grouped is a bond pad. A bond pad usually contains hundreds of vias connected in parallel. Grouping those vias to one single via that has the size of the bond pad does almost not affect the impedance of the corresponding interconnect and hence this will almost not change the substrate frequency response.

    - The edges of the interconnects are aligned (see Figure 2.12). The small metal overlap requires a lot of meshing by the EM simulator. This is very CPU expensive. This small overlap can easily be removed without

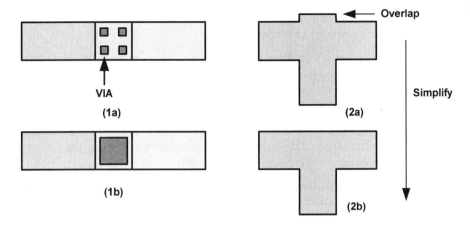

Figure 2.12 (1) The vias are grouped. (2) The edges are aligned.

changing the frequency response. Of course the overlap should not be so large that it determines the capacitance of the interconnect to a large extent.

2. Building the EM environment

The simplified layout is imported into the EM environment, and a simulation model is built in this environment.

- Uniform boxes are added in the EM environment to model the substrate, the p-well, the silicon dioxide between the metal layers, and the air on top of the structures. The size of the boxes is chosen equal to the dimensions of the die or large enough that the electric and magnetic fields are negligible at the boundaries of the EM environment. This ensures that the fields can smoothly radiate away from the EM environment and that there is no interaction that significantly influences the substrate noise propagation. As a rule of thumb, the value of the electric or magnetic fields at the boundaries should be at least three orders of magnitude smaller than in the core of the EM environment.

- The conductivity and the permeability of the different layers are specified. Table 2.5 gives an overview of materials used in this book and their properties. The properties of the substrate layers do not change much with technology scaling. Hence, those values are good starting values for many technology nodes. Note that the use of uniform layers does not require the exact knowledge of the doping profiles.

- Silicon boxes with zero conductivity are added around the n-doped regions to model the depletion layer of the PN junction. The next section is fully devoted to explaining why and how those zero conductivity regions should be chosen such that they correctly model the capacitive behavior of the PN junctions.

- Ports are placed. The location of the ports depends on where the designer requires the knowledge of the S-parameters.

3. Setting up the EM simulations
   The S-parameters are solved in the frequency range of interest for user defined accuracy.

Table 2.5

Materials Used to Characterize the Isolation Structures

Material	Conductivity (S/m)	Relative permittivity
Copper (metal layers)	$5.8 \; 10^7$	1
Silicon dioxide	0	3.7
20 Ωcm Substrate (lightly doped)	5	11.9
P-well (PW)	800	11.9
N-well (NW)	2,300	11.9
T-well (TW)	2,000	11.9
Deep n-well (DNW)	1,200	11.9
P^+/N^+	625,000	11.9

2.4.2   Dealing with N-Doped Regions

An EM simulator can intrinsically not handle n-doped regions such as an n-well. The n-well is usually biased at a constant supply (either V_{DD} or G_{ND}) and forms a PN junction with the p-doped regions that surround the n-well. At the boundaries of the PN junction, there exists a charge-free region called the depletion region. The width of this depletion region is defined by the drift-diffusion equations, and those equations are not solved by the EM simulator. Thus the EM simulator is not capable to handle PN junctions. A possible way out of this problem is to include such n-doped regions in the methodology as follows: the depletion region is approximated by a silicon box of fixed geometry that has zero conductivity (see Figure 2.13). The voltage variations on the power supply are small enough that they do not change the width of the depletion region over time and thus the width of the depletion region can safely be assumed to be fixed. Due to the inserted silicon box, substrate noise can only couple capacitively into the inner part of the box.

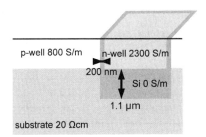

Figure 2.13 The capacitive coupling through the n-well can be approximated by modeling its depletion region with a fixed silicon box.

The value of the PN junction capacitance (C/A in F/m^2) is used to calculate the width (w) of the different depletion regions.

$$w = \frac{\epsilon_r \epsilon_0}{C/A} \qquad (2.34)$$

where ϵ_r reflects the relative permitivity of silicon (11.9) and ϵ_0 is the dielectric permitivity of free space ($8.85 \ 10^{-12}$F/m). The values of the PN junction capacitances can be found in most of the design kits.

Table 2.6 summarizes the different PN junction capacitances that are used to model the depletion regions of the different isolation structures. The capacitances are given for a zero bias of the n-well. When the n-well is biased with the power supply voltage, a different value should be chosen. As a rule of thumb, the value of the PN junction capacitance is approximately four times higher. Note that also the PN junction capacitances are added in this table although the test structure does not contain triple well regions. These values are added because the simulation methodology is capable of handling complex triple well regions.

In the case of the PN junction that is formed between the p-substrate and the n-well, the width of the depletion layer is calculated as:

Table 2.6

PN Junction Capacitances

$C_{TW/NW}$	$1.5\ 10^{-3}$ (F/m^2)
$C_{TW/DNW}$	$1\ 10^{-3}$ (F/m^2)
$C_{PW/NW}$	$5\ 10^{-4}$ (F/m^2)
$C_{NW/SUB} = C_{DNW/SUB}$	$1\ 10^{-4}$ (F/m^2)

$$w = \frac{11.9 \cdot 8.85 10^{-12} F/m}{1 10^{-4}(F/m^2)} \approx 1.1\ \mu m \qquad (2.35)$$

The calculated width of the depletion region at the lateral sides of the n-well is 200 nm. The depletion width at the lateral side of the n-well is much smaller than on the bottom side of the n-well because the p-well has a higher doping concentration than the lightly doped p-substrate. For small n-well regions, substrate noise will couple more through the lateral sides of the n-well than through the bottom side of the n-well. The given widths of the depletion regions are good starting values for many CMOS technologies because the doping concentrations of the p-substrate, the p-well, and the n-well do not change much with scaling.

In the special case of the triple well isolation structure, two silicon boxes with zero conductivity are inserted. The first one models the depletion layer between the p-well/substrate and the n-well. The second one models the depletion layer between the n-well and the P$^+$ substrate contact (see Figure 2.14).

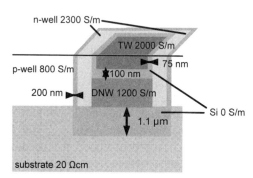

Figure 2.14 Two silicon boxes with zero conductivity model the two PN junctions present in the triple well.

2.4.3   Simulation Setup for the Test Structure

The test structure is described in Section 2.3.1. The corresponding layout is simplified and streamed into the EM environment as explained in the previous section. In
our case, we use HFSS [16] as the EM simulator. The 2D layout is then converted
into a 3D EM simulation model using a technology file. This file contains the thick-
nesses of the different metal and vias. This technology information can easily be
found in the design kit. In this environment a p-well with a conductivity of 800 S/m
and 1.5 μm thick is added on top of a 300 μm thick lightly doped substrate. This
substrate has a resistivity of 20 Ωcm. A silicon dioxide layer with an ϵ_r is added on
top of the p-well layer. The thickness of this layer varies with the technology node
and the number of available metal layers. The thickness is in this case approximately
8 μm. On top of that layer an air box that is 100 μm thick is included.

The simulation is set up as described in the previous section. Two ports are
placed at the two substrate contacts. The p-well, the silicon dioxide, the substrate,
and the air box are added. The PN junction of the n-well and the p-well/substrate is
modeled by a silicon box with zero conductivity.

The HFSS simulation environment is simulated from 50 MHz up to 20 GHz
with a ΔS of 0.001 The simulation takes half an hour on an HP DL145 server.

2.4.4   Comparison

The forward propagation (S_{21}) obtained by both methods match (see Figure 2.15)
with measurements. One can conclude that both methods predict the substrate noise
propagation with the same accuracy.

The electrical field distribution obtained by the HFSS allows to visualize
the substrate noise propagation in a similar way as the SubstrateStorm tool (see
Figure 2.16). However, the HFSS obtains the results about 300 times faster than the
SubstrateStorm tool. It is also important to note that compared with the simulation
tool SubstrateStorm, the HFSS does not need doping profiles.

2.4.5   Conclusions

The FEM simulation tool HFSS is the preferred method to study the isolation structures of the next section. Compared to the FDM simulation tool SubstrateStorm, the
HFSS approach does not need doping profiles. In addition, it has a similar accuracy
and requires less simulation time.

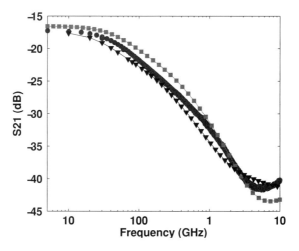

Figure 2.15   Comparison of the forward propagation between measurements and both simulation tools. The ● reflects the measured forward propagation. The □ and the ▽, respectively, denote the simulations performed by SubstrateStorm and HFSS.

2.5   CONCLUSIONS

This chapter studies the propagation of substrate noise on a simple test structure. The study reveals that the propagation of substrate noise in a lightly doped substrate is rather complex. The propagation is determined by layout details such as metal traces and n-well regions. Therefore, it is only possible to provide a theoretical analysis with formulas for very simple geometrical structures. Thus, in order to study complex isolation structures, the designer needs to rely on mesh-based solutions. Two mesh-based solutions are proposed in this chapter: the finite difference method and the finite element method. Both methods are implemented by different simulation tools. The simulation tool SubstrateStorm was used to demonstrate the FDM and the HFSS simulation tool to demonstrate the FEM. Both methods proved to successfully predict the propagation of substrate noise with good accuracy. However the HFSS simulation tool proved to be much faster. Moreover, HFSS does not need the knowledge of the doping profiles. Uniform doping profiles, which are reflected by uniform layers with a predefined conductivity, are sufficient to predict the propagation of substrate noise. Therefore the HFSS will be used in the next chapter to study isolation structures, such as guard rings.

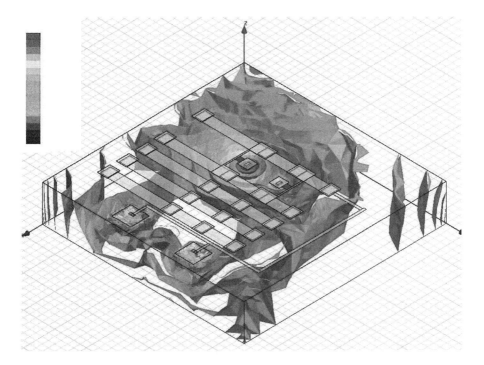

Figure 2.16 Simulated electrical field distribution at 100 MHz. The electrical field distribution provides the same insight in the substrate noise propagation as the equipotential surfaces do. (See color section.)

References

[1] G. Hu and R. Bruce, "A CMOS structure with high latchup holding voltage," *EDL*, Vol. 5, No. 6, 1984, pp. 211–214.

[2] D. Kontos, R. Gauthier, K. Chatty, K. Domanskr, M. Muhammad, C. Seguin, and R. Halbach, "External latchup characteristics under static and transient conditions in advanced bulk CMOS technologies," *Proc. Reliability Physics Symposium 45th Annual. IEEE International*, 2007, pp. 358–363.

[3] T. Blalack, Y. Leclercq, and C. Yue, "On-chip RF isolation techniques," *Proc. Bipolar/BiCMOS Circuits and Technology Meeting the 2002*, September 29–October 1, 2002, pp. 205–211.

[4] D. Su, M. Loinaz, S. Masui, and B. Wooley, "Experimental results and modeling techniques for substrate noise in mixed-signal integrated circuits," *JSSC*, Vol. 28, No. 4, April 1993, pp. 420–430.

[5] C. Soens, C. Crunelle, P. Wambacq, G. Vandersteen, D. Linten, S. Donnay, Y. Rolain, M. Kuijk, and A. Barel, "RF performance degradation due to coupling of digital switching noise in lightly doped substrates," *Proc. Southwest Symposium on Mixed-Signal Design*, February 23–25, 2003, pp. 127–132.

[6] S. Bronckers, C. Soens, G. Van der Plas, G. Vandersteen, and Y. Rolain, "Simulation methodology and experimental verification for the analysis of substrate noise on LC-VCO's," *Proc. Design, Automation & Test in Europe Conference & Exhibition DATE '07*, 2007, pp. 1–6.

[7] S. Ponnapalli, N. Verghese, W. K. Chu, and G. Coram, "Preventing a 'noisequake' [substrate noise analysis]," *IEEE Circuits and Devices Magazine*, Vol. 17, No. 6, 2001, pp. 19–28.

[8] E. Schrik, A. van Genderen, and N. van der Meijs, "Coherent interconnect/substrate modeling using SPACE - an experimental study," *Proc. 33rd Conference on European Solid-State Device Research ESSDERC '03*, 2003, pp. 585–588.

[9] S. Bronckers, K. Scheir, G. Van der Plas, and Y. Rolain, "The impact of substrate noise on a 48-53GHz mm-wave LC-VCO," *Proc. IEEE Topical Meeting on Silicon Monolithic Integrated Circuits in RF Systems SiRF '09*, 2009, pp. 1–4.

[10] K. Joardar, "A simple approach to modeling cross-talk in integrated circuits," *IEEE Journal of Solid States Circuits*, Vol. 29, No. 10, October 1994, pp. 1212–1219.

[11] H. Lin, J. Kuo, R. Sobot, and M. Syrzycki, "Investigation of substrate noise isolation solutions in deep submicron (DSM) CMOS technology," *Proc. Canadian Conference on Electrical and Computer Engineering CCECE 2007*, April 22–26, 2007, pp. 1106–1109.

[12] W. Schoenmaker, P. Meuris, W. Schilders, and D. Ioan, "Modeling of passive-active device interactions," *Proc. 37th European Solid State Device Research Conference ESSDERC 2007*, 2007, pp. 163–166.

[13] A. Afzali-Kusha, M. Nagata, N. Verghese, and D. Allstot, "Substrate noise coupling in SoC design: modeling, avoidance, and validation," *Proceedings of the IEEE*, Vol. 94, No. 12, December 2006, pp. 2109–2138.

[14] *Coupling Wave Solutions*, http://www.cwseda.com.

[15] *Substrate Noise Analysis Cadence*, http://www.cadence.com.

[16] *HFSS*, http://www.ansoft.com/products/hf/hfss/.

[17] N. Verghese, T. Schmerbeck, and D. Allstot, *Simulation Techniques and Solutions for Mixed-Signal Coupling in Integrated Circuits*, Kluwer Academic Publishers, 1995.

[18] C. A. Balanis, *Advanced Engineering Electromagnetics*, Wiley, 1989.

[19] D. S. Burnett, *Finite Element Analysis from Concepts to Applications*, Reading, MA, Addison Wesley Publishing Company, 1987.

[20] H. L. Berkowitz and R. A. Lux, "An efficient integration technique for use in the multilayer analysis of spreading resistance profiles," *Journal Electrochemical Society*, Vol. 128, 1981, pp. 1137–1141.

[21] L. Deferm, C. Claeys, and G. Declerck, "Two- and three-dimensional calculation of substrate resistance," Vol. 35, No. 3, March 1988, pp. 339–352.

[22] R. Piessens, W. Vandervorst, and H. Maes, "Incorporation of a resistivity-dependent contact radius in an accurate integration algorithm for spreading resistance calculations," *Journal Electrochemical Society*, Vol. 130, 1983, p. 468.

[23] B. Stanisic, N. Verghese, R. Rutenbar, L. Carley, and D. Allstot, "Addressing substrate coupling in mixed-mode ICs: simulation and power distribution synthesis," *JSSC*, Vol. 29, No. 3, March 1994, pp. 226–238.

[24] F. Clement, E. Zysman, M. Kayal, and M. Declercq, "LAYIN: toward a global solution for parasitic coupling modeling and visualization," *Proc. Custom Integrated Circuits Conference the IEEE 1994*, May 1–4, 1994, pp. 537–540.

[25] K. Kerns, I. Wemple, and A. Yang, "Stable and efficient reduction of substrate model networks using congruence transforms," *Proc. IEEE/ACM International Conference on Computer-Aided Design ICCAD-95. Digest of Technical Papers*, 1995, pp. 207–214.

[26] M. Pfost, H.-M. Rein, and T. Holzwarth, "Modeling substrate effects in the design of high-speed Si-bipolar ICs," *JSSC*, Vol. 31, No. 10, October 1996, pp. 1493–1501.

[27] X. Aragones and A. Rubio, "Challenges for signal integrity prediction in the next decade," *Materials Science in Semiconductor Processing*, Vol. 6, 2003, pp. 107–117.

[28] M. van Heijningen, J. Compiet, P. Wambacq, S. Donnay, M. Engels, and I. Bolsens, "Analysis and experimental verification of digital substrate noise generation for epi-type substrates," *JSSC*, Vol. 35, No. 7, July 2000, pp. 1002–1008.

[29] G. Van der Plas, C. Soens, G. Vandersteen, P. Wambacq, and S. Donnay, "Analysis of substrate noise propagation in a lightly doped substrate [mixed-signal ICs]," *Proc. Proceeding of the 34th European Solid-State Device Research Conference ESSDERC 2004*, 21–23 September, 2004, pp. 361–364.

[30] M. Pfost and H.-M. Rein, "Modeling and measurement of substrate coupling in Si-bipolar IC's up to 40 GHz," *JSSC*, Vol. 33, No. 4, April 1998, pp. 582–591.

[31] *Spectre RF*, http://www.cadence.com/products/custom_ic/spectrerf.

[32] T. Brandtner and R. Weigel, "Hierarchical simulation of substrate coupling in mixed-signal ICs considering the power supply network," *Proc. Design, Automation and Test in Europe Conference and Exhibition*, March 4–8, 2002, pp. 1028–1032.

Chapter 3

Passive Isolation Structures

3.1 INTRODUCTION

The most popular and straightforward way to shield analog circuits against substrate noise is to use guard rings. Guard rings are passive isolation structures, which prevent substrate noise currents from reaching the analog circuitry [1].

The goal of this chapter is to give the designer a clear understanding about how the substrate isolation is obtained for different types of guard rings. Then, the substrate noise attenuation of the different guard rings are compared against each other and the designer is advised which, when, and how the guard rings should be used. The different types of guard rings that are commonly available to the designer are [2–4]:

- P-well block isolation;

- N-well isolation;

- P$^+$ guard ring shielding;

- Triple well shielding.

In order to have a fair comparison between those different types of guard rings, each type of guard ring is integrated in a separate isolation structure. Each of the isolation structures is simulated by an EM simulator as explained in the previous chapter. The resulting electrical fields provided by the EM simulator gives a deep understanding of how substrate noise propagates through the substrate and how the substrate noise isolation provided by the different isolation structures is achieved.

The designer is taught in this chapter how to interpret the electrical field distribution in complex isolation structures.

Next, it is shown that the proposed simulation methodology can be used to predict the substrate noise isolation between the analog and digital circuitry provided by each of the isolation structures. Therefore the different isolation structures are prototyped and then measured. Comparing the simulation with measurements reveal that the simulation methodology is able to predict the isolation of the different guard ring with an accuracy better than 2 dB.

Then, the attenuation of the different isolation structures is compared. It is shown that P^+ guard ring and the triple well shielding obtain at least 40 dB of isolation compared to the reference structure which has no isolation. The obtained isolation with the P^+ guard ring and the triple well shielding strongly depends on the impedance of their ground interconnects. The designer is instructed on how to choose the isolation structure that best fits his needs.

Finally, the designer is taught how to design an efficient P^+ guard ring. Such a P^+ guard ring provides a good attenuation compared to the other types of guard rings when it is properly designed. The different aspects of the P^+ guard ring design are investigated and then guidelines are formulated for a good P^+ guard ring design.

3.2   OVERVIEW AND DESCRIPTION OF THE DIFFERENT TYPES OF PASSIVE ISOLATION STRUCTURES

The previous chapter pointed out that in the nominal case [see Figure 3.1(a)] most of the substrate noise flows from the aggressor (digital circuitry, power amplifier) to the victim (analog circuitry) through the p-well layer. Consequently there are two approaches to enhance the isolation between the victim and the aggressor. The first approach consists of forcing the current to flow in the lightly doped substrate (20 Ωcm) instead of the conductive p-well (0.2 Ωcm). This can be achieved by inserting p-well block layers [see Figure 3.1(b)] or n-well isolations [see Figure 3.1(c)]. The second approach consists of intercepting the substrate currents and draining them to the off-chip ground through a dedicated ground interconnect. Hence, those substrate currents do not reach the sensitive analog circuitry. This is achieved by placing P^+ guard rings [see Figure 3.1(d)]. Both methods can be combined to obtain an optimal substrate noise isolation. The Triple well shielding is an example of such a combined guard ring (see Figure 3.1(e)). In the triple well shielding, the victim is embedded in a p-doped triple well region which is shielded from the substrate by a deep n-well (DNW)/n-well region. Moreover, a P^+ guard ring surrounds the victim. This P⁺ guard ring intercepts the substrate noise and drains it to the off-chip ground. Note that the triple well shielding can only be used to shield NMOS devices and not an entire analog circuitry consisting of both NMOS and PMOS devices.

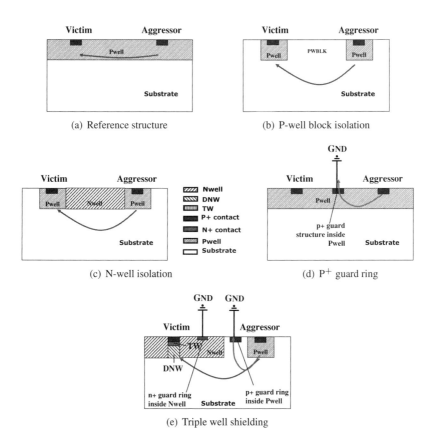

Figure 3.1 (a–e) Cross-section of the mainstream guard rings.

The p-well block isolation, the n-well isolation, P⁺ guard ring, and the triple well shielding guard rings are analyzed in the next section. There, EM simulations are given on how substrate noise propagates through the different guard rings and how the substrate noise isolation is achieved. In order to have a fair comparison of the performance of the different types of guard rings, each type of guard ring is integrated in an isolation structure. It is important to first describe the isolation structures because the substrate noise propagation is influenced by layout details as shown in the previous chapter. This section describes those different isolation structures. All the isolation structures start from the same template layout. After introducing the template layout, it is shown how the different guard rings are integrated in the different isolation structures.

3.2.1   The Template Layout

The previous chapter has shown that the propagation of substrate noise is influenced by layout details like metal interconnects. In order to have a fair comparison of the performance of the different types of guard rings, the different guard rings are integrated in the same template layout. In this way, the performance of the different types of guard rings are influenced by the same layout details.

The template layout consists of a square configuration with 150 μm pitch ground-signal-ground bond pads at each side. Hence, the layout consists of four signal bond pads, which are referred to in Figure 3.2 as ports.

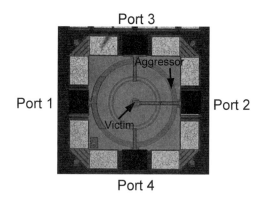

Figure 3.2   Chip photograph of the P$^+$ guard ring isolation structure.

The four ports are defined as follows:

- Port 1 is connected to a substrate contact and acts as the aggressor. This substrate contact is shaped as a P$^+$ ring that surrounds the victim. The aggressor substrate contact has a radius of 140 μm and is 10 μm wide.

- Port 2 is also connected to a substrate contact, which acts as the victim. The victim is shaped as a P$^+$ round contact placed in the center of the structure. The victim substrate contact has a radius of 20 μm.

- Port 3 and Port 4 are left floating or are connected to a guard ring, depending on the isolation structure. For each isolation structure separately, it is explained how Port 3 and Port 4 are connected.

For each isolation structure, a p-well block region is inserted underneath the bond pads such that they do not influence the substrate noise propagation between the aggressor and the victim (see Figure 3.3). Such a p-well block region is defined as a region without the p-well. The interconnect between the victim substrate contact and its corresponding bond pad is shielded with a Metal4 (ME4) ground shield. This ME4 ground shield is connected to a Metal1 (ME1) shield that connects the different ground bond pads. The victim and aggressor substrate contact are connected to the bond pads through a Metal6 (ME6) track. All of those layout details influence the substrate noise propagation.

3.2.2 Integrating the Different Types of Guard Rings

This section describes how the different types of guard rings are embedded in the isolation structures and shows a generic cross-section of the different isolation structures.

1. *P-well block isolation*: A p-well block region is inserted between the two substrate contacts. This p-well block region removes the p-well layer. Only the lightly doped substrate is present in this region. In the p-well block isolation structure, Port 3 and Port 4 are left floating.

2. *N-well isolation*: A n-doped region is inserted between the two substrate contacts. The n-doped region is approximately 1.5 μm thick. Port 3 and Port 4 are left floating in this test structure.

3. *P$^+$ guard ring*: The victim is surrounded by two circular shaped P$^+$-doped regions. Both P$^+$-doped regions are 5 μm wide and 400 nm thick. The inner P$^+$ ring has a radius of 95 μm and the outer P$^+$ ring 125 μm. The inner ring is connected to Port 3 and the outer ring is connected to Port 4.

4. *Triple well shielding*: The victim is surrounded by a circular n-doped region which has a radius of 160 μm and is 5 μm wide. Underneath the victim substrate contact a deep n-doped region (DNW) is inserted. The n-doped region

is connected to Port 3 through N$^+$ contacts. The n-doped region is surrounded by a P$^+$ region which is 5 μm wide and has a radius of 95 μm. The P$^+$ guard ring is connected to Port 4.

3.2.3 Simulation Setup

All the isolation structures are simulated by an EM simulator as explained in the previous chapter. This section explains how the simulation methodology should be applied in order to characterize the isolation structures [5].

The EM simulator *HFSS* [6] is used to analyze the different isolation structures. The simulation setup starts from the layout. The layout is originally drawn in Cadence, using the Virtuoso layout editor. The layout is slightly modified in the Cadence environment to fasten simulations. The vias that connects the different metal layers are grouped and the metal layers are aligned. This modified layout is then imported into the EM environment.

In this environment a p-well with a conductivity of 800 S/m and 1.5 μm thick is added on top of a 300 μm thick lightly doped substrate. This substrate has a resistivity of 20 Ωcm. A silicon dioxide layer with an ϵ_r of 3.7 is added on top of the p-well layer. The thickness of this layer varies with the technology node and the number of available metal layers. The thickness is in this case approximately 8 μm. On top of that layer a 100 μm thick air box is included. A silicon box with zero conductivity is added around the n-doped region. This silicon box models the capacitive coupling through the PN junction.

Then ports are placed. Four lumped ports are placed between the ME1 ground shield and the corresponding signal bond pad.

This simulation environment is simulated from 50 MHz up to 20 GHz with a maximum error in the S-parameters of 0.01. The simulation takes half an hour on a HP DL145 server.

3.3 PREDICTION AND UNDERSTANDING OF GUARD RINGS

This section applies the proposed simulation methodology on the different isolation structures presented in Section 3.2. These are:

- P-well block isolation;
- N-well isolation;
- P$^+$ guard ring;

- Triple well shielding.

The EM simulations allow the visualization of the electric fields in the structure. The electric field distribution allows one to understand how the isolation is obtained for the different types of guard rings. Moreover, the simulations are compared with measurements. As the isolation obtained by the different isolation structures is expressed by the forward propagation S_{21}, the forward propagation S_{21} is measured with a network analyzer. All the isolation structures are measured with a two-port network analyzer from 30 kHz up to 20 GHz. The P$^+$ guard ring and the triple well isolation structure are also measured with a four-port network analyzer from 30 kHz up to 6 GHz. All the performed simulations match with the measurements with an accuracy that is better than 2 dB. Afterwards the performances of the different types of guard rings are compared.

3.3.1 Reference Structure

The reference structure is the structure that does not contain any guard ring (see Figure 3.3). There are two reasons why it is important to analyze the reference structure. First, the reference structure contains the layout details that are also present in the other isolation structures. Hence, this enables the investigations how the layout details influence the substrate noise propagation. Every isolation structure will be influenced in the same way by those layout details. Second, the reference structure allows us to analyze how much isolation the other isolation structure achieved compared with the reference structure, which has no isolation.

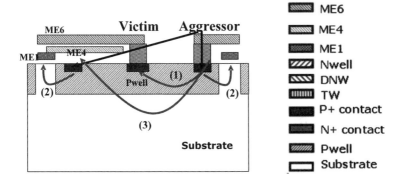

Figure 3.3   Cross-section of the reference structure.

3.3.1.1 Contact-to-Contact Resistance

The measured DC resistance between the substrate contact is 275Ω. Given the circular layout, the resistance between the two substrate contacts can be calculated as:

$$R = \frac{\rho_{pwell}}{2\pi \cdot t_{pwell}} ln(\frac{d+r}{r})$$

(3.1)

where r is the radius of the victim contact, d is the distance between the contacts, ρ_{pwell} is the p-well resistivity and t_{pwell} is the p-well thickness. The calculated value of the resistance between the two substrate contacts is 265Ω, which is close to the measured value. This is consistent with a two dimensional current flow in a thin conducting sheet of material.

The propagation of substrate noise can be visualized by plotting the electric fields in the EM environment [7]. The value of the electric field is two orders of magnitude higher through the p-well layer (0.2 Ωcm) than through the high resistive substrate (20 Ωcm). Remember that the p-well on top of the substrate is two orders of magnitude more conductive than the intrinsic substrate.

3.3.1.2 S-Parameter Measurement

S-parameter measurements and simulations have been performed on the reference structure (see Figure 3.4). At low frequencies substrate noise couples resistively from the aggressor substrate contact toward the victim substrate contact. Hence, the forward propagation is frequency independent. From 30 MHz on, S_{21} decreases with a slope of 10 dB/decade. From 3 GHz on, the slope increases to 20 dB/decade. In the frequency region from 30 kHz up to 3 GHz the error between the measurement and simulations is around 1 dB. For higher frequencies the error is even smaller.

To get a better understanding of the substrate noise propagation, a number of 3D plots are shown in Figure 3.5. Those plots show the electric field distribution on a logarithmic scale. The pseudo coloring in a logarithmic scale can be interpreted as follows: the more red the arrows are, the higher the electric field in that region. A blue arrow corresponds to a low value of the electric field distribution. The propagation of substrate noise can be monitored by the direction of the arrows. In more complex structures, it is more difficult to visualize the propagation of substrate noise with arrows. Therefore, for more complex structures, we will rely on 2D plots of the electric field distribution.

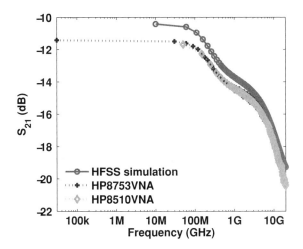

Figure 3.4   Simulated and measured forward propagation of the test structure.

As a first observation of the 3D plots (see Figure 3.5), one can notice that the magnitude of the electric field decreases with increasing frequencies.

At 50 MHz, substrate noise couples directly from the aggressor to the victim [see Figure 3.3(1)]. The electric field distribution at 300 MHz reveals that substrate noise couples capacitively into the ME1 ground shield. Note the change in color in the electric field vectors, located under the signal bond pads between Figure 3.5(a) and Figure 3.5(b). This ME1 shield is picking up substrate noise currents [see Figure 3.3(2)]. Those currents do not reach the victim and hence S_{21} decreases. At 20 GHz the electric field distribution is no longer symmetric [see Figure 3.5(c)]. Note the electric field vectors that point toward this ME4 shield. Here, substrate noise capacitively couples in the ME4 shield [see Figure 3.3(3)]. This ME4 shield shields the interconnect between the victim substrate contact and its corresponding signal bond pad. Hence, with increasing frequencies less substrate noise currents reach the victim substrate contact. Capacitive coupling into the ME1 and the ME4 ground shield explains the increase in the slope of the forward propagation.

3.3.2   P-Well Block Isolation

P-well block isolation means that there is no p-well present in that region (see Figure 3.6). Hence, at DC substrate noise currents are forced to flow from the aggressor

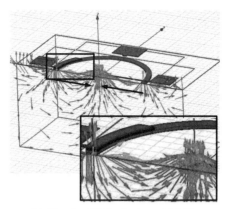

(a) Electric field distribution at 50 MHz

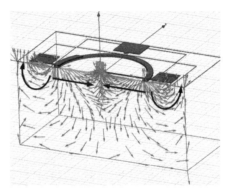

(b) Electric field distribution at 300 MHz

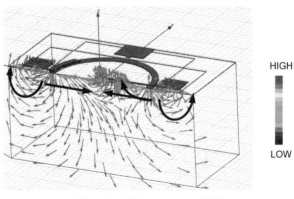

HIGH

LOW

(c) Electric field distribution at 20 GHz

Figure 3.5   (a–c) Electric field distribution at different frequencies. (See color section.)

substrate contact to the victim substrate contact through the bulk substrate. The resistance between the two substrate contacts can be calculated as:

$$R = \frac{\rho_{sub}}{4r} \cdot (1 - \frac{2}{\pi} arcsin(\frac{r}{r+L})) \tag{3.2}$$

with ρ_{sub} the substrate resistivity, r is the radius of the victim substrate contact, and L is the distance to the aggressor. The calculated resistance between the two substrate contacts has a value of $2,272\Omega$ while the measured DC value equals $2,416\Omega$. The error is only 6%. The value of the DC resistance in the case of the p-well block isolation structure is almost one order of magnitude larger than the value of the DC resistance in the reference structure. In the case of the p-well block isolation, the isolation is thus achieved because the current cannot flow through the conductive p-well region anymore.

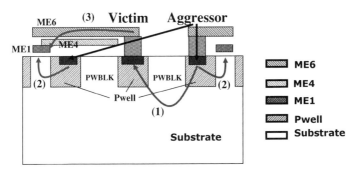

Figure 3.6 Cross-section of p-well block isolation structure.

Figure 3.7 shows the simulated and measured forward propagation (S_{21}) of the p-well block isolation structure. One can immediately see that there is a very good agreement between measurements and simulations. Starting from a few tens of megahertz, the value of S_{21} starts to decrease at a rate of 20 dB/decade. At 3 GHz, S_{21} starts to increase at a rate of 20 dB/decade until it reaches 10 GHz. Starting from 10 GHz, S_{21} starts to decrease again.

Similar to the reference structure studied in Chapter 2, the study of the electrical fields reveals that the decrease in value at low frequencies is due to the capacitive coupling to the ME1 ground shield [see Figure 3.6(1)]. The substrate currents that couple into this ground shield will not reach the victim substrate contact and hence the isolation improves. Starting from 3 GHz, the substrate itself behaves capacitively [see Figure 3.6(2)]. The cutoff frequency of the substrate

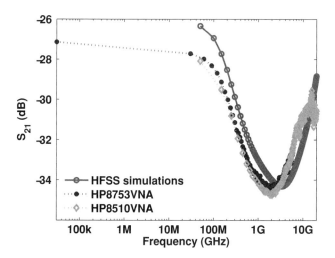

Figure 3.7   Measurement versus simulation for the p-well block isolation.

$(f_{c,sub})$ is lowered from 7 GHz to 3 GHz because the p-well is removed. The capacitive behavior of the substrate provides a lower impedance path than the capacitive coupling to the ME1 shield. This explains the increase in value of S_{21} with 20 dB/decade. Above 20 GHz, S_{21} decreases again in value.

3.3.3   N-Well Isolation

In the case of the n-well isolation, an n-doped region is placed between the two substrate contacts (see Figure 3.8). At DC the substrate currents flow through the bulk substrate from the agressor substrate contact toward the victim substrate contacts. Similar to the case of the p-well block isolation, substrate currents cannot flow through the p-well region and hence a better isolation is obtained compared to the reference structure where this p-well is indeed present.

Figure 3.9 shows the measured and the simulated forward propagation (S_{21}) from 30 kHz up to 20 GHz. The very good agreement between measured and simulated forward propagation shows that the depletion region of the PN junctions can be modeled by inserting zero conductive silicon boxes. Starting from 100 MHz, S_{21} increases with 20 dB/decade. At a frequency of 8 GHz, S_{21} decreases with a rate

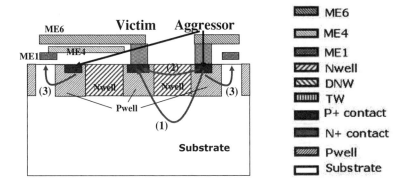

Figure 3.8 Cross-section of n-well isolation structure.

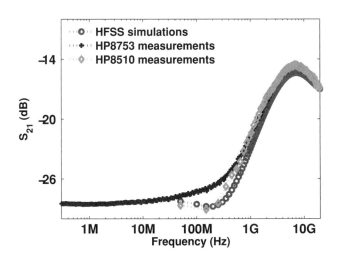

Figure 3.9 Measurement versus simulation for the n-well isolation.

of 20 dB/decade. The behavior is fully explained by inspecting the electric fields in the EM environment.

Plotting the logarithmically scaled electric fields in the n-well for different frequency values reveals that, from 100 MHz on, substrate noise couples capacitively through the n-well isolation [see Figure 3.8(2)]. Figure 3.10 shows the magnitude of the electric fields for different frequencies. Red coloring corresponds to high values for the electric field, while blue coloring reflects low values of the electric fields. From Figure 3.10(a–d), the designer can observe the capacitive coupling through the n-well. The coloring of the electric fields changes from blue toward red with increasing frequencies. At 20 GHz [see Figure 3.10(e)], the n-well colors less red, which corresponds to decreasing values of S_{21}. The inspection of the electric field vector reveals that the dominant substrate couping path at 20 GHz is capacitive toward the ME1 ground shield [see Figure 3.8(2)]. The substrate currents that couple into the ME1 ground shield do not reach the victim substrate contact. The capacitive coupling to the ME1 ground shield enhances the substrate noise isolation.

Note that for frequencies larger than 3 GHz, the substrate noise isolation of the isolating n-well structure is worse than the substrate noise isolation of the reference structure. This is because the n-well is more conductive than the p-well.

3.3.4 P$^+$ Guard Ring Shielding

In the case of the P$^+$ guard ring shielding, two P$^+$-doped ring shaped regions are inserted between the aggressor and the victim substrate contact (see Figure 3.11). The cross-section of this isolation structure is shown in Figure 3.11. Both rings are connected to a dedicated port. Those ports have an impedance of 50Ω. Figure 3.12 shows the measured and simulated forward propagation (S_{21}) from 30 kHz up to 6 GHz. Again, there is very good agreement between measured and simulated results. The error is smaller than 1.5 dB. The EM simulator is able to fully characterize the isolation structure.

Inspecting the electric fields reveal that the P$^+$ guard ring attracts the substrate currents. Figure 3.13 shows the electric fields in the p-well for a frequency of 50 MHz. Note that the p-well is colored in red where the conductive P$^+$ regions are located. The substrate currents are picked up by the guard rings and drained toward the corresponding ports (Port 3 and Port 4). Those currents do not reach the victim substrate contact. Hence, a better isolation is achieved compared to the reference structure. The behavior of the forward propagation (see Figure 3.12) is similar to the behavior of the propagation of the reference structure: starting from a few tens of megahertz, S_{21} decreases due to capacitive coupling to the ME1 shield.

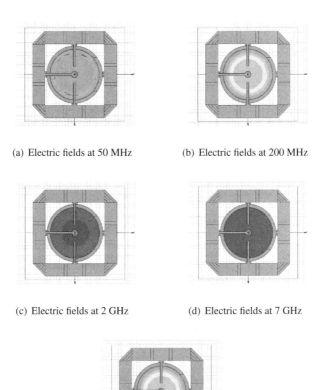

(a) Electric fields at 50 MHz   (b) Electric fields at 200 MHz

(c) Electric fields at 2 GHz   (d) Electric fields at 7 GHz

(e) Electric fields at 20 GHz

Figure 3.10 (a–e) The electric field distribution clearly visualizes the capacitive coupling through the n-well. Red regions correspond to high values of the electric field, blue regions correspond to low values of the electric field. The pseudo-coloring reflects a logarithmic scale. (See color section.)

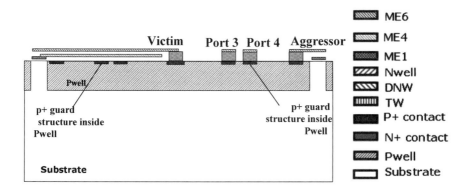

Figure 3.11   Cross-section of the P$^+$ guard ring isolation structure.

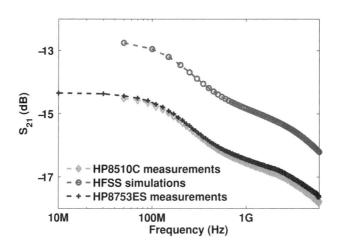

Figure 3.12   Measurement versus simulation for the P$^+$ guard ring.

From 3 GHz on substrate noise also couples into the ME4 shield that shields the interconnect of the victim substrate contact and its corresponding bond pad. This explains the decrease of S_{21} at a higher rate.

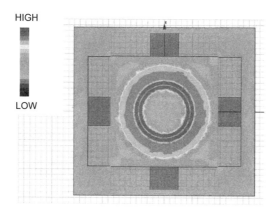

HIGH

LOW

Figure 3.13 Electrical field distribution in the p-well of the P^+ guard ring isolation structure at a frequency of 50 MHz. The red regions correspond to high electrical field values. Blue regions correspond to low electrical field values. The pseudo-coloring reflects a logarithmic scale. (See color section.)

Of course, in a practical context, the P^+ guard rings are never connected to a 50Ω impedance. Instead, the P^+ guard rings are connected to an off-chip ground. In practice, this is usually the ground of the printed circuit board (PCB). The effect of grounding the P^+ guard rings is demonstrated by the following method:

- The resulting S-parameters are imported as an *n-port* in a circuit simulator like SpectreRF [8].

- In the circuit simulator environment, Port 3 and Port 4 of the n-port are connected to the ideal ground.

- An S-parameter analysis is performed on the resulting circuit.

This method will be used in the next section to quantify how low the impedance of the ground interconnect should be chosen. The effect of grounding Port 3 and Port 4 can also be demonstrated by shorting the bond pads of Port 3 and Port 4 in the EM environment to the ME1 ground shield.

Figure 3.14 compares the forward propagation in the case where Port 3 and Port 4 are connected to a 50Ω impedance with the case where both ports are

connected to the ideal ground. Grounding Port 3 and Port 4 improves the isolation
by approximately 20 dB [9].

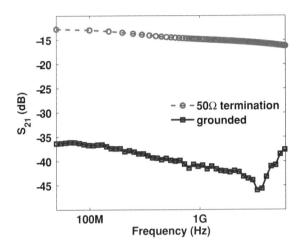

Figure 3.14   Effect of grounding the P$^+$ guard rings.

3.3.5   Triple Well Shielding

In the case of the triple well shielding, the victim substrate contact is fully embedded
in an n-doped region (see Figure 3.15). Furthermore, a P$^+$ guard ring is located
around the victim substrate contact. Here, the P$^+$ guard ring and the n-doped regions
are shorted to the ME1 ground shield of the isolation structure, in order to emulate
the more realistic case where all the guard rings are connected to ground.

Figure 3.16 compares the measured forward propagation (S_{21}) with simulations. There is a very good agreement between both. The error is smaller than 2 dB.
This clearly shows that the proposed methodology is able to predict the substrate
noise propagation of complex isolation structures like the triple well shielding.

Figure 3.16 shows that S_{21} increases at a rate of approximately 20 dB/decade.
This behavior is explained by inspecting the electric field distribution in the isolation
structure. Figure 3.17 shows the electric field distribution of the p-well, n-well,
and the t-well of the isolation structure at a frequency of 50 MHz. Remember that
high values for the electric field distribution correspond with red coloring and low
values with blue coloring. This complex field distribution needs to be interpreted as

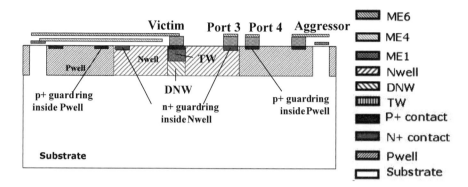

Figure 3.15   Cross-section of the triple well isolation structure.

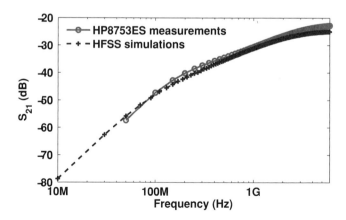

Figure 3.16   Measurements versus simulations for the triple well shielding.

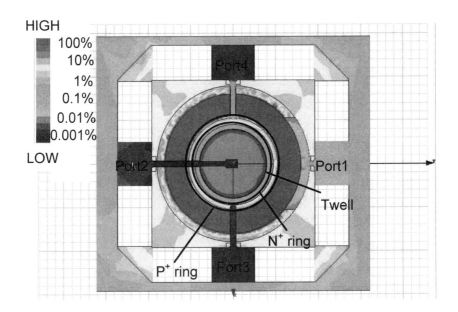

Figure 3.17   Electrical field distribution at 50 MHz. (See color section.)

follows: substrate noise is injected into Port 1, which is connected to the aggressor substrate contact. The substrate currents are then attracted by the P$^+$ guard ring. Note the red region in the inner side of the aggressor substrate contact that spreads out until the P$^+$ guard ring is reached. A part of those currents is picked up by the P$^+$ guard ring and drained toward Port 4. Note that the copper connection of Port 4 colors green. This means that current is transported through this connection. Thus, a first part of the isolation is obtained by the presence of the P$^+$ guard ring. The currents that are not drained by this P$^+$ guard ring propagate further toward the victim substrate contact. Approximately 10% of the currents reach the region between the P$^+$ and the N$^+$ guard ring. Thus approximately 90% of the currents are drained by the P$^+$ guard ring drains. This corresponds to a gained isolation of 20 dB.

The second part of the isolation is obtained by the n-well and the DNW. The n-well and the DNW prevent the substrate current for reaching the victim substrate contact. Only 0.1% of the total current reaches the t-well region. Most of those currents are intercepted by the N$^+$ guard ring and drained through Port 3. Less than 0.01% of the total current reaches the victim substrate contact. The combination of the n-well/DNW and the N$^+$ guard ring enhances the isolation with 30 dB. This explains the isolation enhancement of 50 dB when compared to the reference structure.

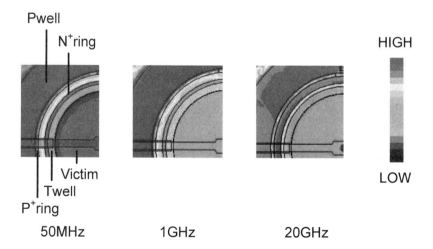

Figure 3.18 Electrical field distribution at 50 MHz, 1 GHz, and 20 GHz. (See color section.)

Most of the isolation is thus obtained by the n-well and DNW region. However, this substrate noise barrier behaves capacitively. This is demonstrated in Figure 3.18 which shows the electric field distribution for three different frequencies, namely: 50 MHz, 1 GHz, and 20 GHz. With increasing frequency the color of the victim substrate contact changes from deep blue toward green. This means that with increasing frequencies, more and more substrate currents reach the victim substrate contact. Substrate noise couples capacitively through the n-well/DNW regions. This explains the increase the S_{21} with increasing frequency. The rate of increase is not exactly 20 dB/decade because different other substrate coupling mechanisms influence the propagation of substrate noise:

- As more currents leaks through the n-well/DNW, more current is drained towards ground via the N^+ guard ring. This can be seen in Figure 3.18. The color of the N^+ guard ring changes from blue to yellow (see the color section).

- Similar to the other isolation structures, at 3 GHz substrate noise couples into the ME1 shield of the isolation structure. Hence, the increase rate of S_{21} diminishes.

- For frequencies larger than 7.5 GHz, the substrate itself behaves capacitively. This effect is clearly included in the simulation model. Comparing the values of the electric field in Figure 3.18 reveals that the value of the electric field in the p-well region between the aggressor substrate contact, and the P^+ guard ring is similar. Both regions are colored in red. However, at a frequency of 20 GHz, the color of this region is orange. Above the cutoff frequency of the substrate, which equals to 7.5 GHz, the bulk substrate behaves capacitively. Less current flows through the p-well and thus less current is attracted and drained by the P^+ guard ring. This effect causes the increase of S_{21} saturates.

3.3.6 Comparison and Conclusion

The previous section discussed, simulated, and experimentally verified the mainstream guard rings that are used by designers. The coupling mechanisms are revealed and a clear understanding is given of how the isolation is obtained for the different types of guard rings. The guard rings can be categorized into power supply free guard rings and guard rings that need a dedicated ground interconnect.

- The p-well block isolation and the n-well isolation guard rings are examples of the first category. Here, the isolation is obtained by forcing the substrate currents to flow in the high resistive substrate (20 Ωcm) instead of the conductive p-well (0.2 Ωcm).

- The P$^+$ guard ring needs a dedicated ground interconnect in order to achieve a good substrate noise isolation. The low-impedance ground interconnect and the dedicated bond pads need to be foreseen in an early stage of the floor planning.

- The isolation strategy of the two guard rings mentioned above can be combined. The triple well isolation structure is an example of such a combined guard ring. Of course, there exists a variety of combined guard rings [10, 11]. The isolation of other types of combined guard rings can easily been obtained with the proposed simulation methodology.

This section first proposes three methods that can be used to obtain the substrate noise isolation for floating and grounded guard rings. Next, the designer is advised as to which guard ring should be used.

3.3.6.1 Methods to Obtain the Substrate Noise Isolation for Floating and Grounded Guard Rings

There are thee different methods to obtain the substrate noise isolation for floating and grounded guard rings:

- Different measurements can be performed to obtain the substrate noise isolation for both cases. In the case where the guard rings are left floating, a two-port VNA measurement between the aggressor and the victim substrate contact is performed. The ports connected to the guard ring are left floating. In the case where the guard rings are grounded, the same VNA measurement is performed when the ports connected to the guard ring are shorted to ground. The easiest way to short those ports to ground is to place an SMA short on top of the probe that connects those ports with the measurement equipment.

- The structures are measured with a four-port network analyzer. The corresponding S-parameters are imported as an n-port in the circuit simulator. In the case where the guard rings are left floating, a resistor with a value of 1 MΩ is placed in the schematic at the guard ring connections [see Figure 3.19(a)]. Then a two-port S-parameter simulation is performed between the aggressor and the victim substrate contact. In the case where the guard rings are grounded, the guard ring connections are connected to the zero potential ground [see Figure 3.19(b)]. Again a two-port S-parameter simulation is performed between the aggressor and the victim substrate contact.

- The substrate noise isolation can be obtained for both cases with EM simulations. In the case where the guard rings are left floating, no ports are placed at the guard ring connections. In the case where the guard rings are grounded, the bond pads of the guard ring connections are shorted with the ground plane of the structure with a connection of perfect conductor material.

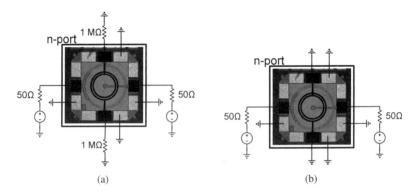

(a)   (b)

Figure 3.19 Method to obtain the substrate noise isolation for floating and grounded guard rings. (a) Schematic of a guard ring with floating guard ring connections. (b) Schematic of a guard ring with grounded guard ring connections.

The three methods provide the same results. However, the second method is the easiest and the fastest to implement. Therefore, the second method is used to compare the substrate noise isolation for the different types of guard rings for both categories:

- Power supply free guard rings;
- Guard rings with a ground connection.

3.3.6.2 Power Supply Free Guard Rings

Figure 3.20 compares the different types of guard rings when the guard ring connections are left floating. One can note that at low frequencies (<10 MHz), the triple well shielding provides the most substrate noise attenuation. Remember that in the case of the triple well shielding, the victim substrate contact is fully shielded by an n-well/DNW. However, already at frequencies as low as 15 MHz, the insertion of a p-well block layer provides more attenuation, even when compared to the n-well isolation. Note that the P^+ guard ring does not provide any isolation when compared to the reference structure. This is logical since the P^+ doped region that attracts the substrate currents is not able to drain those currents to the measurement ground because the ground connections are left floating.

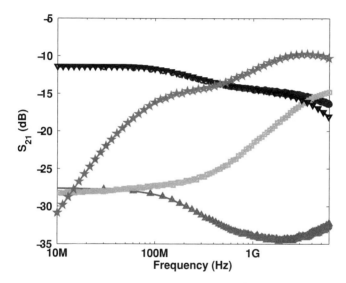

Figure 3.20 Comparison of the studied guard rings when the guard rings are left floating. The ○ reflects the forward propagation of the reference structure, △ the p-well block isolation, □ the isolating n-well, ▽ the P^+ guard ring, and ★ the triple well shielding.

Usually power supply free guard rings can provide a supplementary 15–20 dB of substrate noise isolation when compared to the reference structure.

3.3.6.3 Grounded Guard Rings

Figure 3.21 compares the different types of guard rings when the guard rings are connected to the zero ground potential. Of course, the n-well isolation and the p-well block isolation cannot be connected to a ground connection since they are power supply free guard rings, but it is interesting to compare their substrate noise isolation to the substrate noise isolation of the grounded P⁺ guard ring and the grounded triple well shielding. From Figure 3.21, the triple well shielding provides the most isolation at low frequencies. The isolation is better than 40 dB for frequencies lower than 10 MHz when compared to the reference structure. Starting from 50 MHz, the P⁺ guard ring provides more substrate noise isolation than the triple well shielding. Remember that the P⁺ guard ring structure has two P⁺ guard rings. The triple well shielding has only one P⁺ guard ring. Moreover, the leaking PN junction provides a conductive path for the substrate noise currents through the n-doped regions. This worsens the substrate noise isolation when compared to the P⁺ guarding. The grounded P⁺ guard ring provides at least 10 dB more isolation than the power supply free guard rings.

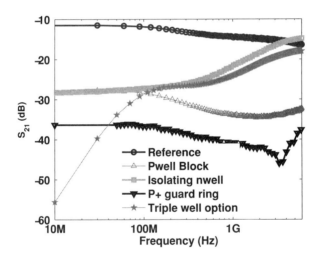

Figure 3.21 Comparison of the studied guard rings when the guard rings are grounded. The ○ reflects the forward propagation of the reference structure, △ the p-well block isolation, □ the n-well isolation, ▽ the P⁺ guard ring, and ★ the triple well shielding.

Grounded guard rings can provide 30 dB or more isolation when compared to the reference structure. The main drawback of the grounded guard rings is the expense of an extra bond pad and the need a low-impedance ground interconnect.

3.4 DESIGN OF AN EFFICIENT P$^+$ GUARD RING

The P$^+$ guard ring is the most popular choice to shield analog circuits from their digital aggressor. The P$^+$ guard ring achieves a good substrate noise isolation compared to other types of guard rings. Only the triple well shielding performs better at low frequencies. However, the triple well shielding needs an additional implant/mask processing steps, which increases cost and process cycle time. P$^+$ guard rings do not need any extra processing steps, and thus they provide a low cost alternative. Moreover, the triple well shielding can only be used to shield NMOS devices and not an entire analog circuit.

The substrate noise isolation provided by P$^+$ guard rings is achieved by attracting the substrate currents and then draining them to the PCB ground through a dedicated ground interconnect. Hence the isolation obtained by such a guard ring depends on three parameters:

- The impedance of the ground interconnect;
- The width of the P$^+$ region of the guard ring;
- The location of the guard ring (i.e., the distance to the victim).

In this section, the designer is taught with measurements how those parameters need to be chosen in order to obtain an efficient P$^+$ guard ring. A guideline is formulated for each of the three parameters that defines the efficiency of the P$^+$ guard ring. Each of those guidelines is only valid when it is used together with the other two guidelines.

3.4.1 Impedance of the Ground Interconnect

The impedance of the ground interconnect determines to a large extent the efficiency of a P$^+$ guard ring. Indeed, substrate currents that are picked up by the P$^+$ region of the guard ring are drained toward PCB ground. Section 3.2 already pointed out that in the extreme case where the guard ring connections are left floating (i.e., the value of the ground resistance is nearly infinite), the P$^+$ guard provides no isolation at all. This section shows with measurements and simulations how low the resistance of the guard rings should be. As an example, the P$^+$ guard ring isolation structure of Section 3.3.4 is chosen. This example consists of two P$^+$ regions, which are 5 μm wide. Each of the guard rings has a dedicated ground connection. The resistance of those ground connections is varied in the following way:

- The structure is measured with a four-port VNA from 50 MHz up to 6 GHz.

- The measured S-parameters are imported as an n-port in the circuit simulator.

- In the environment of the circuit simulator, the value of the resistance of the ground interconnect is increased by adding a resistance at the terminals of the n-port, which corresponds to the guard ring connections.

Figure 3.22 shows the forward propagation of the P^+ guard ring at a frequency of 1 MHz for different values of the resistance of the ground interconnect. This figure clearly shows that the forward propagation S_{21} increases with increasing values of the resistance of the ground interconnect. Hence, the isolation ($1/S_{21}$) degrades for larger values of the resistance of the ground interconnect. For a value of 3Ω the isolation is already degraded with 3 dB. Hence, as a first guideline, the resistance of the ground interconnect should be smaller than 1Ω.

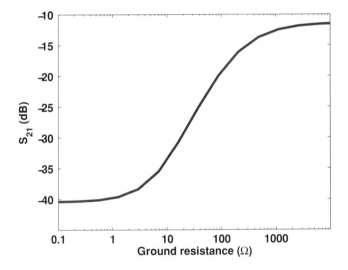

Figure 3.22   Forward propagation in the P^+ guard ring isolation structure in function of the resistance of the ground interconnect for a frequency of 1 MHz.

In a real design, the P^+ guard ring is connected to the PCB through a bond wire or a solder bump in the case of flip-chip. Those connections to the PCB behave inductively. For increasing frequencies, the inductive part of the impedance of the ground interconnect gains importance compared with the resistance of the ground interconnect and the bond wire. Figure 3.23 shows the forward propagation with and without bond wire. The inductance of the bond wire is chosen to be 1 nH. This corresponds to a bond wire that is approximately 1 mm long. From this figure, it can be seen that the inductance of the bond wire limits the frequency range where the P^+ guard ring is effective to 1 GHz. This frequency range can be extended by using multiple bond wires at the expense of multiple bond pads or by using active noise suppression guard rings at the expense of extra power consumption [12, 13].

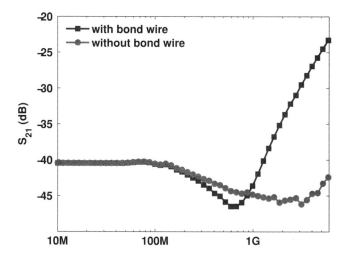

Figure 3.23 Influence of a bond wire of 1 nH on the forward propagation of the P^+ guard ring isolation structure.

3.4.2 Width of the P^+ Guard Ring

In this section, the concept of a sizeable guard ring is introduced. A sizeable guard ring is a P^+ guard ring where the width can be varied without reprocessing the guard ring with a different size. The latter is very expensive. The isolation is measured for the different widths of the sizeable guard ring. Measurements show that the isolation of the guard ring against substrate noise does not increase linearly with the width of the P^+ layer. The isolation saturates with the width of the P^+ layer.

3.4.2.1   A Sizeable Guard Ring

This section introduces the concept of a sizeable guard ring [14]. The goal of a sizeable guard ring is to help the designer determine the optimal width of the P^+ guard ring. Such a sizeable guard ring consists of a certain number of P^+ guard rings. The rings are connected to each other by NMOS devices that act as switches. The outer ring is connected to the PCB ground by a bond wire. The switches allow the P^+ region to be enlarged, hence, the guard ring. Opening such a switch is a very good approximation to a reduction of the size of the ring as Section 3.3.4 showed that a grounded P^+ guard ring provides 30 dB more noise suppression then when it is left floating. A floating guard ring will not drain the intercepted substrate noise. The on-resistance of the switches is less than 1Ω, which guarantees a short circuit between the guard rings and a low impedant connection to the PCB ground. The off resistance of the switches is higher than $10\ k\Omega$. The disadvantage of this setup is that the high off capacitance of the switches results in a de facto connection of the guard rings at high frequencies even if the switch is off. This limits the usability of the sizeable guard ring to an upper frequency of 30 MHz. This is not an issue, since the goal of this experiment is to determine the optimal guard ring width.

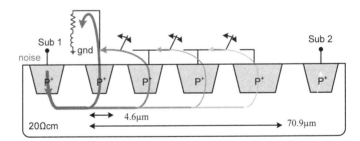

Figure 3.24   Schematic of the sizeable guard ring.

Such a sizeable guard ring is designed in a 0.13 μm CMOS technology. The sizeable guard ring can be enlarged in four steps from 4.6 μm to 71 μm. To this end, different guard rings of respectively 4.6 μm, 9.3 μm, 18 μm, and 39 μm are placed around a very sensitive RF circuit, in this case an LC-VCO (Figure 3.25) (see Figure 3.24). This emulates a more realistic case where the sizes of the guard ring are chosen in order to protect a real analog circuit. Two substrate contacts are foreseen. One substrate contact is placed inside the ring and the other substrate contact outside the ring. Both the substrate contacts have a size of 115 μm $\times$ 58 μm. In this way, the forward propagation (S_{21}) between the two substrate contacts is representative for the efficiency of the P$^+$ guard ring. The impact of substrate noise on the LC-VCO, which lies inside the guard rings, is studied in Chapter 5.

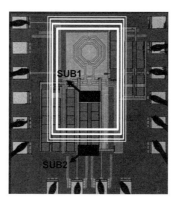

Figure 3.25   An LC-VCO is surrounded by four guard rings connected to each other via switches.

3.4.2.2   Effectiveness of the P$^+$ Guard Ring

Measurement of the S-parameters for the different widths of the sizeable guard ring will provide insight to the analog designer in how the width of the P$^+$ layer determines the efficiency of the P$^+$ guard ring.

The S-parameters are measured from a substrate contact that is realized outside the ring (SUB2) to another substrate contact that is placed inside the ring (SUB1) (Figure 3.25). Figure 3.26 shows the S-parameters in a frequency band ranging from 30 kHz up to 300 MHz for a guard ring width of 32 μm. As expected, the S-parameters are symmetrical for this passive structure. There is no influence of the switches for frequencies lower than 10 MHz. Starting from 10 MHz, the off capacitance of the switches determines the frequency behavior of S_{21}. S_{11} and S_{22} are frequency independent over the whole band as expected.

The S-parameters are measured for four different widths of the sizeable guard ring. Figure 3.27(a) shows the transfer function that is computed between the two substrate contacts based on the S-parameters measurements. Widening the guard ring from 4.6 μm to 71 μm enhances the isolation between the two substrate contacts by 10 dB. The measurements indicate that the isolation of the guard ring for widths above 16 μm does not increase linearly with the width of the guard ring but rather with the logarithm of the width as shown in Figure 3.27(b). As a second

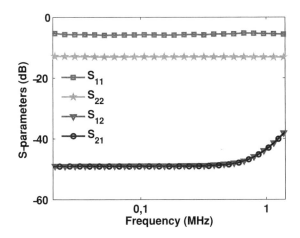

Figure 3.26 Measured S-parameters for a guard ring width of 32 μm. Note that S_{21} and S_{12} are on top of each other.

guideline, the width of the P$^+$ layer should be no more than 10–20 μm large. By making the guard ring larger, not much extra isolation will be gained, but a lot of expensive silicon area will be lost.

3.4.3 Distance to the Victim

This section gives insight into how the isolation provided by a P$^+$ guard ring is affected by the location of the guard ring. Therefore, three isolation structures are considered where the location and the size of the guard ring are varied (see Figure 3.28).

- Structure I considers a small victim, which is protected by a small P$^+$ guard ring that is located close to the victim substrate contact [see Figure 3.28(a)].

- Structure II considers a small victim, which is protected by a large P$^+$ guard ring that is located relatively far from the victim substrate contact [see Figure 3.28(b)].

- Structure III considers a large victim, which is protected by a large P$^+$ guard ring that is located close to the victim substrate contact [see Figure 3.28(b)].

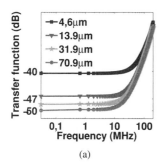
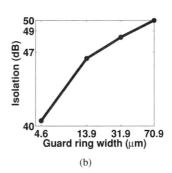

(a)   (b)

Figure 3.27   (a–b) The isolation of the P$^+$ guard ring saturates with the guard ring width.

Table 3.1

Description of the Structures Geometry

Structure	I	II	III
Aggressor radius	160 μm	160 μm	160 μm
Victim radius	20 μm	20 μm	80 μm
d_1	30 μm	90 μm	90 μm
d_2	10 μm	70 μm	10 μm

The distances between the victim and the aggressor substrate contact and the P$^+$ guard ring is summarized in Table 3.1. Distance d_1 reflects the distance between the center of the victim and the P$^+$ guard ring and hence the size of the guard ring. Distance d_2 reflects the distance between the victim substrate contact and the P$^+$ guard ring.

Figure 3.29 shows the measured forward propagation S$_{21}$ for the three structures from 10 MHz up to 1 GHz when the P$^+$ guard rings are connected to a zero ground potential. Comparing structure I with structure II reveals that the combination of making the P$^+$ guard ring smaller and placing it closer to the victim does not improve the isolation a lot. The forward propagation S$_{21}$ of structure I is on average only 1 dB worse than in structure II. On the other hand, in structure III, the size of the victim substrate contact is increased such that the victim substrate contact is located close to the P$^+$ guard ring. This improves the isolation by more than 7 dB compared to the forward propagation of structure II. In order to give a deep understanding about how the location of the guard ring affects the isolation, a lumped

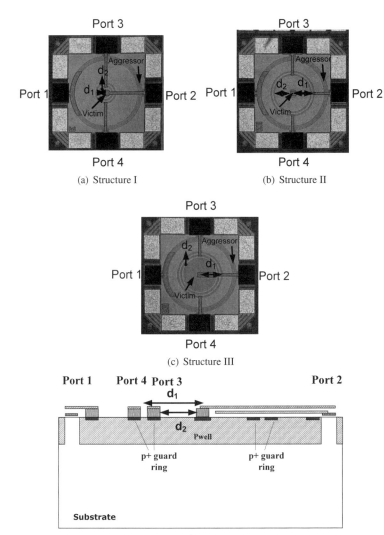

(a) Structure I   (b) Structure II

(c) Structure III

(d) Cross-section of the P$^+$ guard ring isolation structure.

Figure 3.28   (a–d) Influence distance.

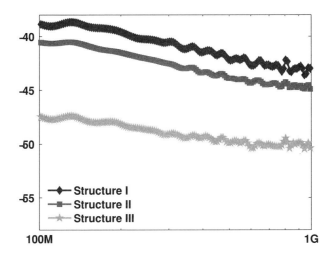

Figure 3.29 Forward propagation of the three structures.

network representation is build for the different structures. First, it is explained how to build such a lumped network from S-parameters. Then the method is applied on the different structures.

3.4.3.1   Method for Deriving Lumped Networks from S-Parameters

A lumped network clearly represents how the lumped elements from the network influence the behavior of the structure in general. The lumped network is built from the S-parameters. One can use either simulated or measured S-parameters. First, the S-parameters are converted into Y-parameters. The Y-parameters reflect the different admittances between the nodes. In this case, how to build such a network that consists of only resistors is shown. This network is only valid for low frequencies where capacitive and inductive effects can be neglected. The general resistive network that corresponds to a Y-parameter matrix is a network where each node is connected to any other node with a resistor. The value of every resistor separately corresponds to the real part of a Y-parameter of the Y-parameter matrix. To be more concrete, this method is applied to an example, which is structure I [see Figure 3.28(a)]. The S-parameters at 50 MHz are:

$$\mathbf{S} = \begin{pmatrix} S_{11} & S_{12} \\ S_{21} & S_{22} \end{pmatrix} = \begin{pmatrix} 4 \cdot 10^{-3} - 3 \cdot 10^{-6} \cdot i & 0.1 - 2 \cdot 10^{-6} \cdot i \\ 0.1 - 2 \cdot 10^{-6} \cdot i & -0.4 - 3 \cdot 10^{-7} \cdot i \end{pmatrix}$$

In this case, the imaginary part of the S-parameter matrix is negligible with respect to the real part. Consequently the structure can be described at 50 MHz by a resistive lumped network. The real part of the S-parameter matrix is then converted into a Y-parameter matrix:

$$\mathbf{Y} = \begin{pmatrix} Y_{11} & Y_{12} \\ Y_{21} & Y_{22} \end{pmatrix} = \begin{pmatrix} 0.02 & -7 \cdot 10^{-4} \\ -7 \cdot 10^{-4} & 0.05 \end{pmatrix}$$

This Y-parameter matrix corresponds to the lumped network that is shown in Figure 3.30. Now, the designer can map the values of the different elements on the Y-parameters of the Y-parameter matrix. The resulting network describes the structure for low frequencies. One can extend the lumped network in a similar way with capacitors and inductors.

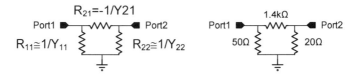

Figure 3.30   From the Y-parameters, a lumped network is built for structure I.

3.4.3.2   Understanding How the Location Affects the Isolation of the Guard Ring

The above method is also used to build a resistive lumped network for structure II and structure III. Their lumped networks are shown in Figure 3.31. In order to have a good isolation from Port 1 to Port 2, the value of R_{21} should be as high as possible and the value of R_{11} and R_{22} should be as low as possible. The value R_{21} of structure I and structure II is 1.4 kΩ and 4.3 kΩ, respectively. Note that the value of R_{21} in the lumped network of structure II is thus almost three times larger than the one in structure I. This is explained because the P$^+$ guard ring of structure II has almost five times the surface of the P$^+$ guard ring of structure I. Remember from the previous section that the isolation of the guard ring saturates with the width (or the surface) of the P$^+$ guard ring. Hence, an increase of a factor of five in the guard ring surface only results in an increase of a factor of three of R_{21}.

However, in structure I, the P$^+$ guard ring is located much closer to the victim substrate contact than in structure II. The proximity of the location of the guard ring to the victim substrate contact mostly affects R_{22} in the lumped network shown in Figure 3.31. The value of R_{22} in structure I and structure II is, respectively, 20Ω and 208Ω. The value of R_{22} in structure I is thus 10 times smaller than in structure II. Hence, the substrate and the p-well underneath the victim substrate contact are better silenced in structure I than in structure II. Thus, as a third guideline, the closer that the P$^+$ guard ring is located to the victim, the better the substrate noise isolation.

Structure I benefits from the proximity to the victim substrate contact and structure II benefits from the large P$^+$ guard ring. The proximity to the victim and the large guard ring influences the isolation of the structures almost with the same amount. Hence, the difference in isolation between structure I and structure II is only 1 dB.

Structure III benefits both from the proximity of the P$^+$ guard ring ($R_{22} = 77\Omega$) to the victim substrate contact and a large P$^+$ guard ring ($R_{21} = 10.9$ kΩ). Hence the isolation of structure III is 7 dB better than the isolation of structure II and 8 dB better than the isolation of structure I.

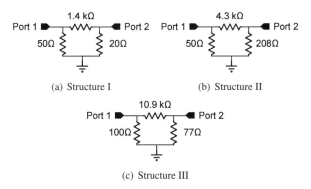

(a) Structure I   (b) Structure II

(c) Structure III

Figure 3.31 (a–c) Influence distance.

3.4.4 Guidelines for Good P$^+$ Guard Ring Design

The P$^+$ guard ring is the preferred guard ring for a designer. There are two reasons for this:

- Good efficiency compared to other guard ring types;

- Low cost because a P^+ guard ring does not need an extra processing step.

A P^+ guard ring consists of a P^+ region. This P^+ region attracts the substrate currents and drains them toward PCB ground. To design such a P^+ guard ring, three guidelines can be formulated for the designer. It is important to note that those guidelines are only valid when they are all used together.

1. The ground interconnect should be as low an impedance as possible. Starting from 1Ω for the resistance of the ground interconnect, the isolation provided by the guard ring starts to degrade. The inductive behavior of the connection from the chip through the PCB increases the interconnection impedance and hence limits the frequency range where the P^+ guard ring is efficient.

2. Do not make the P^+ guard ring too large. The isolation provided by the guard ring saturates with the guard ring width. As a guideline, the width of the guard ring should be around 10 and 20 μm.

3. Place the P^+ guard ring as close as possible to the analog circuitry. The P^+ guard ring will silence the substrate and the p-well under the analog circuitry.

3.5   CONCLUSIONS

This chapter discusses the different types of guard rings that are mostly available for the designer. It is shown that EM simulations are able to predict the isolation provided by those different guard rings with a good accuracy. Moreover, EM simulations can give a deep understanding of how the different types of guard rings achieve substrate noise isolation. The guard rings are categorized into power supply free guard rings and guard rings that need a ground interconnect.

- *Power supply free guard rings* obtain substrate noise isolation by forcing the substrate currents to flow into the high resistive substrate.

- *Guard rings that need a ground interconnect* attract the substrate currents and drain them toward PCB ground.

Comparing the different types of guard rings with each other reveals that if the designer does not want to spend a dedicated ground interconnect and a corresponding bond pad, the p-well block isolation guard ring provides the most isolation. If a dedicated ground interconnect and a corresponding bond pad is available, the triple well shielding provides the most isolation. However, the triple well shielding requires an extra mask, which increases cost. The P^+ guard ring thus provides a low-cost alternative. Moreover the triple well shielding can only be used to shield NMOS devices. The P^+ guard ring can be used to shield an entire analog circuit.

In this chapter, the designer is also taught how to design a good P^+ guard ring. To that end, three guidelines are formulated. Each of the guidelines are established based on dedicated measurement experiments. The guidelines are related to the impedance of the ground interconnect of the P^+ guard ring, its size, and its location:

- The impedance of the ground interconnect should be lower than $1\,\Omega$.

- The isolation of the P^+ guard ring saturates with the width of the P^+ region. Based on measurements, the guard ring should be around 10–$20\ \mu$m.

- The P^+ guard ring should be placed as close as possible to the analog circuitry.

In Chapter 2 and in this chapter, the designer is taught how substrate noise propagates through the substrate, how the propagation is affected by layout details, and how the analog circuitries can be shielded with passive isolation structures. The next chapters focus on the impact of substrate noise on analog circuits. Chapter 4 studies the impact of substrate noise on a single active device. Chapters 5 and 6 investigate the impact of substrate noise on individual analog/RF circuits, and finally Chapter 7 examines the impact of substrate noise on complete analog/RF systems.

References

[1] X. Aragones, J. L. Gonzalez, and A. Rubio, *Analysis and Solutions for Switching Noise Coupling in Mixed-Signal ICs*,Boston,   Kluwer Academic Publishers, 1999.

[2] D. Kosaka and M. Nagata, "Equivalent circuit modeling of guard ring structures for evaluation of substrate crosstalk isolation," *Proc. Asia and South Pacific Conference on Design Automation*, January 24–27, 2006, p. 6.

[3] W.-K. Yeh, S.-M. Chen, and Y.-K. Fang, "Substrate noise-coupling characterization and efficient suppression in CMOS technology," *ED*, Vol. 51, No. 5, May 2004, pp. 817–819.

[4] K. Joardar, "A simple approach to modeling cross-talk in integrated circuits," *IEEE Journal of Solid States Circuits*, Vol. 29, No. 10, October 1994, pp. 1212–1219.

[5] R. Vinella, G. Van der Plas, C. Soens, M. Rizzi, and B. Castagnolo, "Substrate noise isolation experiments in a 0.18 μm 1P6M triple-well CMOS process on a lightly doped substrate," *Proc. IEEE Instrumentation and Measurement Technology*, 2007, pp. 1–6.

[6] *HFSS*, http://www.ansoft.com/products/hf/hfss/.

[7] D. White and M. Stowell, "Full-wave simulation of electromagnetic coupling effects in RF and mixed-signal ICs using a time-domain finite-element method," *IEEE Transaction on Microwave Theory and Techniques*, Vol. 52, No. 5, 2004, pp. 1404–1413.

 [8] *Spectre RF*, http://www.cadence.com/products/custom_ic/spectrerf.

 [9] A. Pun, T. Yeung, J. Lau, J. Clement, and D. Su, "Substrate noise coupling through planar spiral inductor," *JSSC*, Vol. 33, No. 6, June 1998, pp. 877–884.

[10] T.-L. Hsu, Y.-C. Chen, H.-C. Tseng, V. Liang, and J. Jan, "Psub guard ring design and modeling for the purpose of substrate noise isolation in the SOC era," *EDL*, Vol. 26, No. 9, September 2005, pp. 693–695.

[11] J. Lee, F. Wang, A. Phanse, and L. Smith, "Substrate cross talk noise characterization and prevention in 0.35μ CMOS technology," *Proc. Custom Integrated Circuits the IEEE 1999*, May 16–19, 1999, pp. 479–482.

[12] H.-M. Chao, W.-S. Wuen, and K.-A. Wen, "An active guarding circuit design for wideband substrate noise suppression," *MTT*, Vol. 56, No. 11, November 2008, pp. 2609–2619.

[13] W. Winkler and F. Herzel, "Active substrate noise suppression in mixed-signal circuits using on-chip driven guard rings," *Proc. CICC Custom Integrated Circuits Conference the IEEE 2000*, May 21–24, 2000, pp. 357–360.

[14] S. Bronckers, G. Vandersteen, G. Van der Plas, and Y. Rolain, "On the P+ guard ring sizing strategy to shield against substrate noise," *Proc. IEEE Radio Frequency Integrated Circuits (RFIC) Symposium*, June 3–5, 2007, pp. 753–756.

Chapter 4

Noise Coupling in Active Devices

4.1 INTRODUCTION

Transceivers are mainly built from active devices (i.e., transistors). Before the impact of substrate noise can be investigated on analog/RF circuits and even on transceivers, one should start to qualitatively analyze the impact of substrate noise at the transistor level. The goal of this qualitative analysis is to get a better understanding on how substrate noise couples into a transistor and what is required to model the impact of substrate noise on an analog design. To that end, the methodology that is proposed in Chapter 2 to predict the impact of substrate noise on passive isolation structures needs to be refined such that it can handle active devices. One big challenge here is that those active devices are the result of complex fabrication steps, partly relying on the actual doping profiles of the die. Those doping profiles are usually kept confidential by the foundry and therefore cannot be incorporated as public knowledge in the substrate noise prediction. A possible way out of this problem is to note that the impact of the doping profiles on the transistors behavior is well included in the RF model. This RF model is validated up to high frequencies. In the proposed methodology the actual active devices are replaced by their RF model and thus there is no need for the explicit knowledge of the doping profiles. Similar to Chapter 2, the methodology describes the passive part (i.e., the substrate and the interconnects) with a finite element model. This chapter leads the reader through the complete process of how to correctly set up the simulation environment to characterize a single active device. Next, it will be shown that this methodology is able to accurately characterize the substrate together with the active devices that are integrated on this substrate. To this end, the impact of substrate noise on a single transistor is predicted. This case study gives the analog designer the different paths where substrate noise can couple in a transistor and how the adjacent interconnects determine the nature of the dominant coupling mechanism. The next chapter will extend the applicability of the methodology to CMOS analog/RF circuits.

4.2   SUBSTRATE NOISE IMPACT ON ANALOG DESIGN

This section provides a theoretical framework to describe the substrate noise impact on analog design at the transistor level.

Substrate noise has an influence on the drain current, I_d, through the bulk effect and through ground bounce. The bulk effect is defined here as any perturbation on the bulk terminal of the transistor. Further, ground bounce is defined as any perturbation on the ground interconnect. For the sake of qualitative reasoning, we assume that ground bounce directly affects the source terminal of the transistor. This is true since in most of the cases the transistor is connected with its source terminal to the ground interconnect. The drain current is given by the following equation [1]:

$$I_d = \frac{\mu C_{ox}}{2} \frac{W}{L} (V_{gs} - V_t)^2 \tag{4.1}$$

and the threshold voltage V_t equals:

$$V_t = V_{to} + \gamma \cdot \left(\sqrt{\phi + V_{SB}} - \sqrt{\phi} \right) \tag{4.2}$$

where μ is the mobility factor, C_{ox} the oxide capacitance, W the width of the transistor, L the length of the transistor, γ the body-effect coefficient, ϕ is the inverse surface potential, and V_{to} is the threshold voltage for $V_{SB} = 0$.

A Taylor expansion of (4.2) shows that V_t to first order depends linearly on V_{SB}:

$$V_t = V_{to} + \frac{1}{2} \cdot \frac{\gamma}{\sqrt{\phi}} \cdot V_{SB} \tag{4.3}$$

From (4.1) and (4.3), one can notice that the drain current depends both on the voltage on the source terminal and on the bulk terminal. The drain current depends on the substrate voltage through V_{SB}. Further it is obvious if the circuit suffers from ground bounce caused by substrate noise, the drain current through V_{GS} and V_{SB} will be affected.

The drain current, I_d, is primarily defined by the nominal operation conditions of the transistor. Hence the total drain current can be defined as the sum of the nominal drain current and the variation of the drain current caused by substrate noise:

$$I_d(tot) = I_d(nom) + \Delta I_d \qquad (4.4)$$

where $\Delta I_d <<< I_d(nom)$ because substrate noise is a small signal phenomenon. ΔI_d can be written as:

$$\Delta I_d = \frac{\partial I_d}{\partial V_{GS}} \cdot \Delta V_{GS} + \frac{\partial I_d}{\partial V_{BS}} \cdot \Delta V_{BS} \qquad (4.5)$$

or [1]:

$$\Delta I_d = g_m \cdot \Delta V_{GS} + g_{mb} \cdot \Delta V_{BS} \qquad (4.6)$$

where g_m is the transconductance and g_{mb} the body transconductance of the transistor. Remember that ΔI_d is the unwanted variation of the drain current caused by substrate noise. Hence, the gate voltage V_g is equal to zero, and (4.6) can be rewritten as:

$$\Delta I_d = (g_m + g_{mb}) \cdot \Delta V_S + g_{mb} \cdot \Delta V_B \approx g_m \cdot \Delta V_S + g_{mb} \cdot \Delta V_B \qquad (4.7)$$

because g_{mb} is usually one order of magnitude smaller than g_m. From (4.7) one can notice that any perturbation at the source terminal is amplified with the transconductance, g_m, of the transistor. A perturbation at the bulk terminal of the transistor is amplified with the body transconductance, g_{mb}, of the transistor [2]. From this qualitative reasoning, it cannot be determined whether ground bounce or the bulk effect dominates because the perturbation of the source and bulk terminal depends on the transfer function from the digital circuitry (in this case, the substrate contact) toward the different terminals of the transistor. An EM simulator is needed to calculate the amount of substrate noise that reaches the terminals of the transistor. The corresponding changes in the drain current of the transistor can be calculated by the transistor model equations. The next section explains how the methodology of Chapter 3 can be refined such that it can handle transistors. The refined methodology shall be used in Section 4.5 to reveal whether ground bounce or the bulk effect dominates in (4.7).

From the previous equations the following conclusions can be drawn about the transistor sensitivity to substrate noise:

- Equation (4.1) shows that the drain current is function of both V_{GS} and V_{BS}. This means that the variation of the drain current of a transistor will be affected at the same time by the bulk effect and ground bounce. The relation between the bulk effect and ground bounce will be explained in Section 4.5.

- The substrate noise sensitivity of a single transistor is reduced for devices with a large V_{GS}, which is typically the case for transistors used as a current source, for example. A transistor with a small V_{GS} on the other hand will be more sensitive to substrate noise, which is typically the case for input transistors of a circuit that are usually designed for high g_m values.

In this qualitative reasoning it is implicitly assumed that the main coupling mechanisms are either through the bulk or to the source of the transistor. Another coupling mechanism is a capacitive feedthrough of substrate noise from the bulk node to the source, drain, and gate terminals of the transistor via the different junction capacitances. Section 4.5 will show that this only occurs at higher frequencies.

4.3   IMPACT SIMULATION METHODOLOGY

In the previous chapter the EM simulator proved to be a very powerful tool to analyze the propagation of substrate noise. Such an EM simulator only solves the Maxwell equations and not the drift-diffusion equations, which describe the behavior of the active devices. Hence, EM simulations are not capable of modeling active devices. Moreover, in order to solve the drift-diffusion equations, the simulator would require the in-depth knowledge of the doping profiles used to fabricate the active devices. A possible alternative is to use the device knowledge implicitly rather than explicitly. The behavior of the active devices is included into the RF models and hence the usage of the RF models avoids the need to characterize the active devices all over again. In the methodology proposed here, the active and the passive part of the design are modeled separately. The active devices are described by the respective RF models. The passive part (i.e., the substrate and the interconnects) are described by a finite element model. Consequently the substrate and the interconnects are described by small-signal S-parameters. The S-parameters obtained from the EM simulation are then cosimulated with the RF models of the active devices with a circuit simulator. The resulting waveforms on the different terminals of the simulation model gives the designer insight in the different substrate noise coupling mechanisms. The methodology proposed here contains the best of both worlds: the excellent ability of the EM field solver to describe the frequency dynamics of the substrate and the interconnects and the accurate behavioral prediction of device level models are combined to solve the substrate noise characterization problem of active devices. The implementation of this cooperation between two different types of simulations is, as could be expected, not a perfectly straightforward process. However, a clever combination of the capabilities of the different tools enables the designer to obtain a very accurate prediction of the substrate noise impact on active devices for an acceptable effort. A block diagram representation of the methodology is given in Figure 4.1. The different simulations that are needed to predict the impact of substrate noise on active devices are described in more detail and the user interaction that is required is explained.

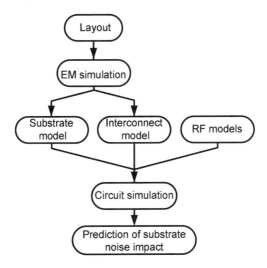

Figure 4.1 Impact simulation approach.

4.3.1 EM Simulation

The goal of the EM simulation is to extract a nonparametric model of the interconnects and the substrate. The EM simulation starts from the layout. The layout is simplified to speed up simulations. The idea used here relies on the following simple observation: the EM simulation of the complete circuit is so CPU expensive that it is not possible. Fortunately, this level of detail is not required. If the layout is simplified, it is noted that this only marginally influences the frequency response. The layout should not be oversimplified because this can change the frequency response dramatically and cause a false prediction of the substrate noise impact on active devices. The goal here is to design a controlled heuristic procedure to simplify the layout as much as possible without jeopardizing the accuracy of the frequency response.

- The different vias that connect the different metal layers are grouped to one single via (see Figure 4.2). The impedance of the interconnect with and without grouped vias should not change drastically. A typical example where the vias can safely be grouped is a bond pad. A bond pad usually contains hundreds of vias connected in parallel. Grouping those vias into a single one with the size of the bond pad almost does not affect the impedance of the corresponding interconnect and hence this will almost not change the substrate frequency response.

- The edges of the interconnects are aligned (see Figure 4.2). The small metal overlap requires a lot of meshing by the EM simulator. This is very CPU expensive. This small overlap can easily been removed without changing the frequency response. Of course the overlap may not be so large that it determines the capacitance of the interconnect to a large extent.

- The ground plane that is slotted to satisfy the DRC rules is filled with metal.

  Those simplifications can be automated.

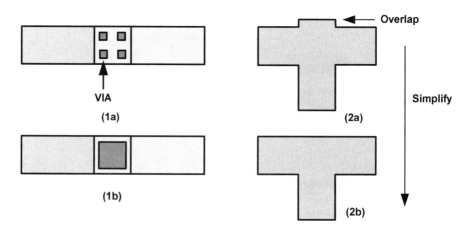

Figure 4.2   (1) The vias are grouped. (2) The edges are aligned.

Furthermore, the transistors are removed from the layout. The substrate contacts, which set the potential of the bulk, are kept. In the case of a multifinger transistor, the different bulk regions, which are located underneath the gate region, are merged into one large bulk region (see Figure 4.3). This means that the bulk is considered equipotential underneath the transistor. This is a safe assumption since the transistors are very small compared to the surface of the IC. Moreover, this assumption is also made by other commercial available tools [3, 4]. Next, this simplified layout is exported as a gds II file. This gds II file is then imported into the EM environment.

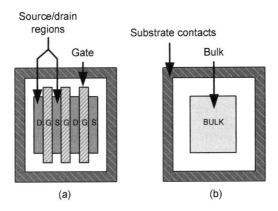

Figure 4.3 (a) Typical layout of a RF-transistor. (b) The substrate contacts are kept and the different bulk regions underneath the gates are merged.

The transistor, which is a four-port device, is modeled in the EM environment by three ports. The drain, gate, and bulk connections of the transistor are replaced by a port that is referred to the source connection of the transistor (see Figure 4.4). Furthermore, additional ports are placed at the position of the different external connections (i.e., at the bond pads of the IC). Appendix B clearly explains how a transistor can be modeled by three ports.

After the port placement, the substrate, the p-well, the silicon dioxide, and the air box on top of the IC are included in the EM environment. The size of the boxes is chosen to be equal to the dimensions of the die or sufficiently large enough that the electric and magnetic fields are small enough at the boundaries of the EM environment. It this way it is ensured that the fields can smoothly radiate away from the EM environment and that there is thus no interaction that significantly influences the substrate noise propagation. As a rule of thumb, the value of the electric or magnetic fields should be at least three orders of magnitude smaller than in the core of the EM environment.

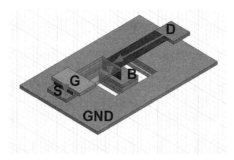

Figure 4.4 Port placement at the connections of the transistor. Here, the drain, gate, and bulk connections of the transistor are referred to the source of the transistor.

Then the simulator is started and the Maxwell's equations are solved at the boundaries of the different ports. The result of the EM simulation is an $n \times n$ matrix of S-parameters where n reflects the number of ports.

4.3.2   Circuit Simulation

A simulation model that fully characterizes the active device is constructed. The substrate and the interconnects are represented by the S-parameter box resulting from the EM simulation. This S-parameter box is included into the netlist together with the RF models of the different devices. It is important to note that the n-port S-parameter box is used here as a $2n$-terminal S-parameter box, where one port consists of a set of two matching terminals. In this case the port conditions still remain satisfied. Appendix B discusses the port conditions. The analog designer can apply any analysis on this simulation model. The corresponding waveforms of the analysis will give insight into how and where substrate noise couples into the active devices.

4.4   TRANSISTOR TEST BENCH

First, it will be shown that the methodology is able to accurately characterize a test structure consisting of a common-source transistor. The designer is taught how to set up the EM simulation and how to correctly connect the S-parameter box to the transistor. In the next section, the impact of substrate noise on a single transistor is investigated.

4.4.1 Description of the Transistor Under Test

The transistor test bench consists of a common-source NMOS transistor realized in a 90 nm CMOS technology on a lightly doped substrate (20 Ωcm). The transistor is quite large in order to obtain a good SNR. Therefore it counts 40 fingers, each of which have a width of 1 μm and a minimal length. The drain and gate of the transistor can be accessed via external bond pads. The source and the bulk are connected to the ground bond pads through a large ground plane. The layout is constructed to be perfectly symmetrical (see Figure 4.5).

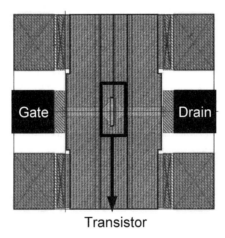

Figure 4.5 Layout of the transistor test bench.

4.4.2 Modeling the Transistor Test Bench

The proposed methodology consists of two simulations:

1. An EM simulation that characterizes the substrate and the interconnects;

2. A circuit simulation to combine the previous simulation with the active transistor models, to obtain a simulation model that fully characterizes the transistor test bench.

   This section explains how to set up both simulations in the case of the transistor test bench.

4.4.2.1 EM Simulation

The goal of the EM simulation is to build a model for the substrate and the interconnects. The EM simulation starts from the layout. The simplifications mentioned
above are performed: In this case the vias are grouped, the metal traces are aligned,
and the slotted ground plane is filled with metal. Furthermore, the transistor is removed from the layout. Its substrate contacts are kept and the different bulk regions
are merged into one bulk region as explained in Section 4.3.1. The remaining geometry is then streamed in the EM simulator (HFSS [5]). In the HFSS environment,
the transistor leads are replaced by three lumped ports. The drain, gate, and bulk
of the transistor are referred to the source of the transistor as shown in Figure 4.6.
Further, a lumped port is placed at each of the two external connections. The HFSS
environment counts therefore five ports (see Figure 4.6).

Once the ports are placed, a substrate of 20 Ωcm, a p-well of 800 S/m, the
silicon dioxide with an ϵ_r of 3.7, and an air box on top of the IC are included in the
EM environment. Since this test bench is meant to be validated with measurements
with on-wafer GSG (ground-signal-ground) probes, it is important to short the
ground bond pads located on both sides of the Signal bond pad (see Figures 4.6 and
4.7). This short circuit is performed in HFSS with a connection made out of perfect
conductor material. This models the fact that in the measurements, the current can
flow back to the measurement equipment through the two ground pins of the GSG
probe.

This HFSS environment is simulated from DC up to 110 GHz with a minimum solved frequency of 1 GHz and a maximum error of the S-parameters of
0.006. The HFSS simulation takes 47 minutes on an HP DL145 server. Figure 4.7
shows the electric field distribution in the test structure. One can clearly see that the
electric field distribution is symmetrical as expected.

4.4.2.2 Circuit Simulation

A circuit simulation model is then constructed that fully characterizes the transistor
test bench. The interconnects, the substrate, and the passive components are represented by an S-parameter box resulting from the HFSS simulation. The transistor is
represented by its RF model and is connected to the S-parameter box. The netlist
below clearly shows how the S-parameter box is connected to the transistor and the
external ports. Those external ports reflect the ports of the measurement equipment.
The ports of the measurement equipment are connected to the positive terminal of
the first two ports of the five-port obtained by the EM simulation. The negative

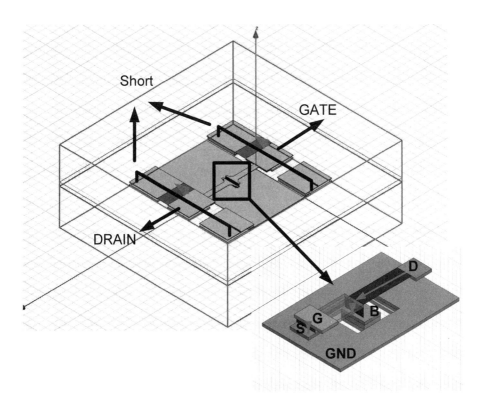

Figure 4.6 View of the HFSS environment. The lumped ports are indicated using arrows. The transistor is simplified to clearly show how the designer should place the ports at the leads of the transistor.

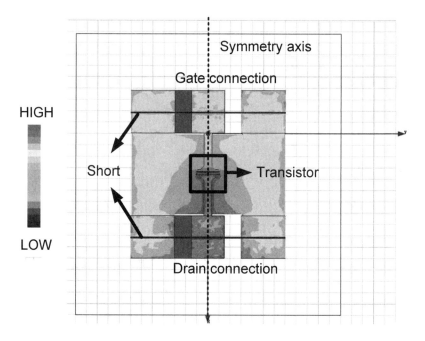

Figure 4.7   The electrical field distribution present in the transistor test bench. The field distribution is symmetrical. (See color section.)

terminal of the first two ports is connected to the ideal ground because it is assumed that the ground of the measurement equipment is ideal. The transistor is connected to the last three ports of the five-port. The positive terminals of ports 3, 4, and 5 are respectively connected to the drain (D), gate (G), and bulk (B) of the transistor while the negative terminal is connected to the source (S) of the transistor (see Figure 4.8).

```
% Netlist of the Transistor test bench

% external connections
PORTGATE (Gate 0) port r=50 dc=0.5 type=dc
PORTDRAIN (Drain 0) port r=50 dc=1 type=sine

% transistor
NMOSname (D G S B) NMOS parameters l=, w=, nf= ...

% substrate + interconnects
Sparam ( Drain 0 Gate 0 D S G S B S) nport file="name"
```

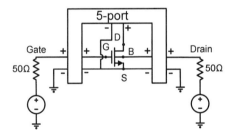

Figure 4.8 Schematic of the transistor test bench. The five-port resulting from the EM simulations is properly connected to the transistor and the measurement ports.

On this simulation model an S-parameter analysis is performed with Spectr-eRF [6]. The analysis takes less than 1 minute on an HP Proliant DL145G1 platform. When an S-parameter analysis is performed, a DC analysis is performed first. The DC operating point of the transistor is calculated based on the DC results of the S-parameter box obtained by the HFSS simulation. HFSS extrapolates the DC operating points from its minimal solved frequency, which is a minimum of 50 MHz. The HFSS tool sometimes suffers from the fact that this extrapolation is not always performed correctly. However this problem can easily be circumvented by placing an ideal bias tee at one of the leads of the transistor. In this way the designer can set the correct DC potential. The correct potential of the bias tee can, for example, be obtained from a limited parasitic extraction of the test structure.

4.4.3   Experimental Validation

To show that this methodology truly can handle active devices, experimental validation is indispensable. Measurements are the only way to verify if the performed simulations are correct. Therefore the transistor test bench is measured using a VNA. The test bench is accessed with on-wafer probes and calibrated up to the probe tips. The transistor is biased as in simulation: the drain is biased at 1V and the gate at 0.5V. Then, the S-parameters of the test bench are measured with the network analyzer from 1 GHz up to 110 GHz. Figure 4.9 shows that there is a very good agreement between measured and simulated S-parameters, both in amplitude and in phase.

The mean error is calculated as in [7]:

$$Error = \frac{100}{N} \left[\sum_{i=1}^{2} \sum_{j=1}^{2} \frac{|\Re(S_{i,j,meas}) - \Re(S_{i,j,sim})|^2}{|S_{i,j,meas}^2|} \right.$$
$$\left. + \sum_{i=1}^{2} \sum_{j=1}^{2} \frac{|\Im(S_{i,j,meas}) - \Im(S_{i,j,sim})|^2}{|S_{i,j,meas}^2|} \right] \quad (4.8)$$

N reflects the number of measurement points. The indices *meas* and *sim* denote the measured and simulated S-parameters, respectively. In the case of the transistor test bench, the error is less than 3%. Therefore, it can be concluded that this methodolgy can handle active devices.

4.5   SUBSTRATE NOISE COUPLING MECHANISMS IN A TRANSISTOR

Up until now it has been shown that the proposed methodology is able to accurately characterize a transistor and its interconnects. This section also includes the influence of the substrate, predicts the impact of substrate noise on active devices, and reveals the different coupling mechanisms.

Generally speaking, at low frequencies where capacitive and inductive effects can be neglected, substrate noise can only couple resistively into the transistor. There are two possible paths for this resistive coupling:

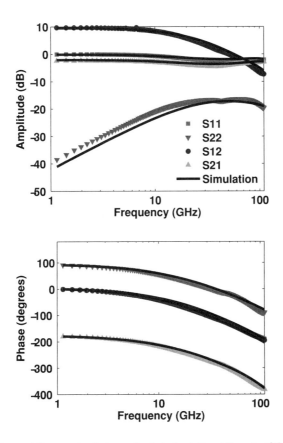

Figure 4.9 Measured S-parameters between the drain (port 1) and the gate of the transistor (port 2). The □ reflects S_{11}, O S_{12}, △ S_{12}, and ▽ S_{22}. The lines are the corresponding simulations.

- *Resistively into the bulk of the transistor:* Substrate noise then modulates the bulk voltage of the transistor.

- *Resistively into the P+ guard ring of the transistor:* Here, substrate noise couples into the ground interconnect of the transistor and causes ground bounce.

Measurements cannot always reveal which of the above mentioned coupling paths dominates. This is because only the external connections can be sensed during measurements. The internal nodes like for example the bulk of the transistor are not available during measurements. In this case, simulations can provide supplementary information about the nature of the coupling mechanisms. Simulations can also pinpoint which circuit parameters influence the substrate noise behavior before the IC is fabricated. The designer can then tune those parameters such that the IC is less sensitive to substrate noise perturbations.

Therefore the proposed methodology is used to predict the impact of substrate noise on active devices. Here, an EM simulation characterizes the substrate and the interconnects. The resulting EM model is connected to the RF models of the transistors. The resulting simulation model is analyzed with a circuit simulator. The designer can extract supplementary information about the substrate noise coupling mechanisms from both the EM and the circuit simulation:

- *Propagation through the substrate*: The electric field distribution provided by the EM simulator gives the designer insight into how substrate noise propagates through the substrate. Analyzing the substrate noise propagation gives a good indication how and where substrate noise couples into the transistor and which circuit parameters influence the substrate noise coupling mechanism.

- *Propagation through the transistor*: The circuit simulator provides the voltage and currents at the different nodes of the simulation model. Analyzing the voltages and currents and the different nodes give insight into how substrate noise propagates through the transistor. Moreover the designer can now modify the parameters of the simulation model. Modifying the parameters of the simulation model give the designer insight into which parameter influences the substrate noise coupling mechanism.

4.5.1 Analyzing the Different Coupling Mechanisms in a Transistor

It was shown earlier that at sufficiently low frequencies, where capacitive and inductive effects can be negated, substrate noise propagates resistively through the substrate. At such low frequencies substrate noise will couple resistively into the transistor. Substrate noise can couple resistively either into the bulk of the transistor or the P+ guard ring that surrounds the transistor. This section analyzes the different substrate noise coupling mechanisms in the case where the transistor is common-source coupled. Similar expressions can be found for other transistor configurations. In the next section, this analysis will be used to model the substrate noise coupling mechanisms on a real-life example.

The proposed theoretical analysis is based on the lumped equivalent network that is shown in Figure 4.10. This lumped network is used to analyze the low frequency substrate noise coupling mechanisms in a common-source coupled transistor. For the sake of simplicity, substrate noise originates from one single node. Hence, the voltage transfer function from the substrate contact (SUB) to the drain (D) of the common-source coupled transistor is given by:

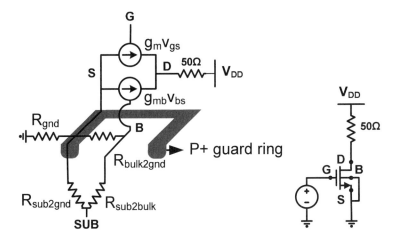

Figure 4.10 Lumped network that is used to model the different substrate noise coupling mechanisms in a common-source coupled transistor. S, D, B, and G reflect the source, drain, bulk and gate terminals of the transistors. SUB indicates the equivalent node from where the substrate noise originates.

$$TF_{sub2drain} \approx -\frac{R_{bulk2gnd} + R_{gnd}}{R_{sub2bulk} + R_{bulk2gnd} + R_{gnd}} \cdot g_{mb} \cdot R_{load}$$

$$+ \frac{R_{gnd}}{R_{sub2gnd} + R_{gnd}} \cdot g_m \cdot R_{load} \qquad (4.9)$$

where

- $R_{bulk2gnd}$ is the bulk resistance.
- $R_{sub2bulk}$ is the resistance between the SUB node and the bulk of the transistor.
- g_{mb} is the body transconductance of the transistor.
- R_{load} is the load resistance that is connected to the drain of the transistor.
- R_{gnd} is the resistance of the ground interconnect.
- $R_{sub2gnd}$ is the resistance between the SUB node and the ground interconnect of the transistor.
- g_m is the transconductance of the transistor.

Equation (4.10) can be simplified by the a priori knowledge that $R_{sub2bulk} >> R_{bulk2gnd}$, $R_{sub2gnd} >> R_{gnd}$, and $R_{bulk2gnd} >> R_{gnd}$:

$$TF_{sub2drain} \approx -\frac{R_{bulk2gnd}}{R_{sub2bulk}} \cdot g_{mb} \cdot R_{load} + \frac{R_{gnd}}{R_{sub2gnd}} \cdot g_m \cdot R_{load} \qquad (4.10)$$

Similar expressions can be derived for other transistor configurations. This is left as an exercise to the reader. One can identify two parts in (4.10). The first part is given by:

$$-\frac{R_{bulk2gnd}}{R_{sub2bulk}} \cdot g_{mb} \cdot R_{load} \qquad (4.11)$$

Equation (4.11) represents the part of the transfer function that reflects the bulk effect. The bulk effect is dominant in the case where the value of the impedance of the ground interconnect is very small compared to the bulk resistance ($R_{bulk2gnd}$).

The second part of (4.10) is given by:

$$\frac{R_{gnd}}{R_{sub2gnd}} \cdot g_m \cdot R_{load} \qquad (4.12)$$

This term models the *ground bounce* as a coupling mechanism. Note that ground bounce gets more important than the bulk effect for higher values of the impedance of the ground interconnect. When ground bounce dominates, substrate noise couples directly into the P+ guard ring of the transistor and causes a voltage variation on the internal ground interconnect node.

It is important to note in (4.10) that the two coupling mechanisms have an opposite sign. Hence, there exists an optimum where both coupling mechanisms are canceled and the isolation is optimal. The only circuit element in (4.10) that can be easily adjusted by the designer is the resistance of the ground interconnect R_{gnd}.

The optimal value for the resistance of the ground interconnect can be calculated as:

$$R_{gnd} = \frac{R_{bulk2gnd}}{R_{sub2bulk}} \cdot g_{mb} \cdot \frac{R_{sub2gnd}}{g_m} \tag{4.13}$$

In order to quantify the value of the optimal ground resistance and to check the validity of the derived equations, a dedicated test structure is built. This test structure consists of a common-source coupled transistor designed in a 130 nm CMOS technology. The next section describes the test structure and presents the corresponding measurements.

4.5.2 Description and Measurement of the Device Under Test

4.5.2.1 Description of the Device Under Test

In order to study the different coupling mechanisms that are present for a single transistor, a simple test structure is designed. The structure consists of eight parallel connected common-source NMOS transistors configured in a 130 nm triple well technology. The dimensions of the transistors are chosen large to obtain a good signal-to-noise ratio. Consequently they each count 16 fingers, which have a minimal length and are 3.6 μm wide. The gate of the transistor is ESD protected for measurement purposes. The p-type ESD diodes are embedded in an n-well and the n-type diodes in a triple well. Hence substrate noise can only couple capacitively through the PN junctions. Thus the ESD diodes will not influence the substrate noise coupling mechanisms into a transistor at low frequencies.

The V_{DD}, which biases the n-well of the triple well transistor (see Figure 4.11), is decoupled with MIM-capacitors. The corresponding chip photograph is shown in Figure 4.12.

A dedicated substrate contact with a size of 114 μm by 58 μm is placed next to the transistor. A substrate contact acts as a resistive connection between the measurement equipment and the substrate. Hence, such a substrate contact can be driven by a source to replace the digital switching noise in this experiment in a very controlled way.

Two versions of this test structure are processed. In the first version the NMOS transistor is shielded by a triple well. The second version is the twin well version

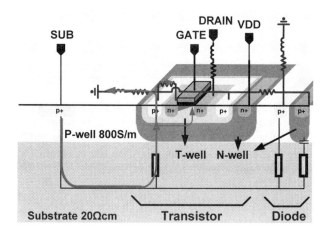

Figure 4.11   Cross-section of the triple well structure. The impact of substrate currents are indicated by the arrows comprising both ground bounce and bulk-source effects.

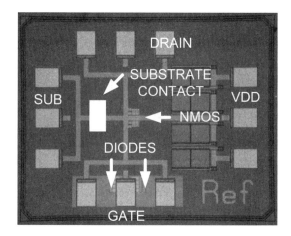

Figure 4.12   Die photograph of the test structure.

where this extra well is not present. This allows one to assess the effectiveness of the triple well option to shield an NMOS transistor.

4.5.2.2 Measurement of the Device Under Test

The device under test is measured using on-wafer probes. The gate of the transistor is biased at 0.6V. The drain and the V_{DD} terminal are biased at a power supply of 1.2V. The S-parameters of the two versions of the test structures are measured with a four-port VNA from 1 MHz up to 6 GHz. The calibration is performed up to the probe tips.

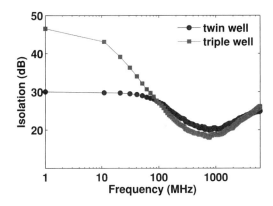

Figure 4.13   The triple well option enhances the substrate noise isolation with 18 dB at low frequencies when compared to the twin well version.

Figure 4.13 shows the isolation between the substrate contact (port 1) and the drain of the transistor (port 2) for the two versions of the test structure. The isolation is here defined as $1/S_{21}$. The triple well enhances the isolation with 18 dB at low frequencies when compared to the twin well version. However, starting from 5 MHz, this isolation starts to decrease because the shielding of the n-well starts to leak due to its junction capacitance. For noise frequencies larger than 300 MHz there is no difference in isolation between the two versions.

For both versions the phase shift of S_{21} is 180 degrees at 1 MHz, which means that the bulk effect dominates in both cases. Since the low frequency coupling mechanisms are the same for both versions, the next section will focus on the twin-well version of the test structure only.

Measurements combined with the analysis presented in the previous section are able to pinpoint which substrate noise coupling mechanism is dominant. However, these measurements can neither reveal nor quantify the parameters that influence the coupling mechanisms. This is because the internal nodes of the test structure, like for example the bulk below the channel of the gates, cannot be accessed during measurements. In that case simulations can provide more information about the different coupling mechanisms as will be shown next.

4.5.3   Modeling Different Substrate Noise Coupling Mechanisms

In order to reveal and to quantify the parameters that determine the dominant coupling mechanisms, a simulation model is built. To that end, the proposed methodology of Section 4.3 is used. This methodology uses an EM simulator like HFSS [5] to extract a model of the substrate, the passive devices, and the interconnects. The active devices like the transistor and the ESD diodes are represented by their RF model as was explained before.

4.5.3.1   EM Simulation

The EM simulation starts from the layout of the test structure. The transistors are removed from the layout as was explained before. The P^+ guard ring surrounding each of the 8 RF-transistors is kept in layout. Each transistor counts 16 fingers and thus has 16 separate bulk regions. Those different bulk regions are merged into one bulk region like in [3]. Therefore the bulk resistance will be slightly underestimated.

Further, the layout is simplified to speed up simulations. The simplifications consist in this particular case in grouping the vias that connect the different metal layers and aligning the different metal layers. This modified layout is then streamed into the EM simulator. In this case HFSS was used. A substrate of 20 Ωcm, a p-well of approximately 800 S/m, the silicon dioxide with an ϵ_r of 3.7, and an air box on top of the circuit are included in the HFSS environment. Also the shallow trench isolation (STI) regions are included in the EM environment (see Figure 4.14). Those STI regions determine to a large extent the value of the bulk resistance [8]. Here, the bulk resistance is defined as the resistance between the bulk of the transistor and the P+ guard ring that surrounds the RF transistor.

Next, the simulation is set up and ports are placed. These ports form the interface between the EM model and the external world. The interface between the interconnects and the transistor is formed by three ports. A port is placed at the internal gate, drain, and bulk connection, which is referred to the internal

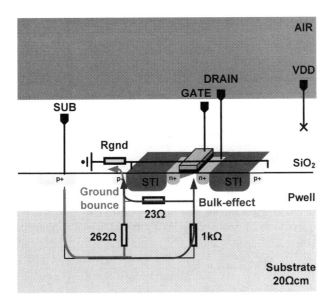

Figure 4.14 Cross-section of the twin-well NMOS transistor. For simplicity only the STI regions between the source/drain regions and the P+ guard ring are drawn.

source connection of the transistor. This source connection is part of the ground interconnect of the test structure. Further ports are placed at the four external connections of the test structure.

The PN junctions of the different wells of the ESD diodes are modeled by inserting a zero conductive silicon region between the different wells to represent the depleted region. In this way substrate noise can only propagate capacitively through the different junctions. This is a known technique, which is also used in [9].

Consequently, the HFSS environment counts seven ports. The S-parameters are solved from DC up to 6 GHz with a minimum solved frequency of 50 MHz and a maximum S-parameter error of 0.01. The HFSS simulation takes about 1.5 hours on an HP DL145 server.

4.5.3.2   Circuit Simulation

A simulation model is constructed that fully characterizes the test structure. As explained before, the interconnects, the substrate, and the passive components are represented by an S-parameter box resulting from the HFSS simulation. The transistor and the ESD diodes are represented by their RF model and are connected to the S-parameter box. If the RF model of the transistor contains a dedicated network for the substrate, this substrate network should be removed. On this simulation model an S-parameter analysis is performed with SpectreRF [6]. The analysis takes less than 1 minute on an HP Proliant DL145G1 platform.

4.5.3.3   Experimental Validation

It is important to validate the simulation model with measurements. This is the only way to check whether the simulation model incorporates the dominant coupling mechanisms. Figure 4.15 shows that there is a very good agreement between measurements and simulations both in amplitude and in phase. The mean error, calculated as in [7], is 1 dB. At low frequencies the amplitude error of S_{21} is 2 dB due to the underestimation of the bulk resistance. The simulated bulk resistance has a value of 19Ω. The value of the bulk resistance of the RF model is approximately 23Ω.

It can be concluded that the model incorporates the dominant coupling mechanisms. The internal nodes of the test structure can now be accessed in simulation. Hence, the designer can modify the parameters of the simulation model and can reveal which parameters influence the coupling mechanism.

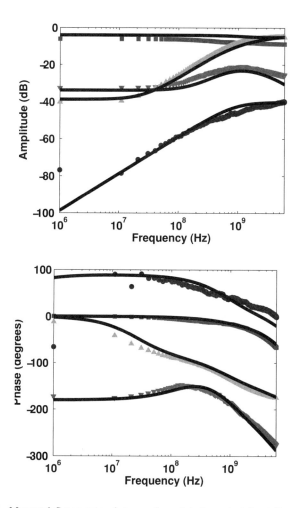

Figure 4.15 Measured S-parameters between the substrate contact (port 1) and the drain of the transistor (port 2). The □ reflects S_{11}, • S_{21}, △ S_{12}, and ▽ S_{22}. The black lines are the corresponding simulations. The measurement noise floor is around -80 dB.

4.5.4   Quantifying the Different Substrate Noise Coupling Mechanisms

This section quantifies the different substrate noise coupling mechanisms. First, the electrical field distribution provided by the EM simulation is used to study the propagation of substrate noise and to determine the different substrate noise entry points. Then, the parameters that define the different substrate noise coupling mechanisms are quantified based on the analysis that is built in the previous section. The values of the parameters are derived from the S-parameters that are obtained from the EM simulation.

4.5.4.1   Propagation of Substrate Noise

The propagation of substrate noise can be visualized by plotting the electric fields in the HFSS environment [10]. Figure 4.16 shows the electric fields at 100 MHz. It can be seen that substrate noise couples into the P+ guard rings that surround the transistors and couples into the ground interconnect of the transistors (the dark region at the P+ guard rings of the eight parallel connected transistors). The electric field distribution clearly shows that the ground interconnect plays a dominant role in the impact of substrate noise. Of course, the electric field distribution cannot reveal the different coupling mechanisms in a transistor since the transistor itself is not included in the EM simulation. Therefore, the complete simulation model that comprises the transistor models needs to be studied.

4.5.4.2   Analyzing the Substrate Noise Coupling Mechanisms

In the previous section an analysis was built to study the different coupling mechanisms in a source-coupled transistor. This analysis is based on a simple lumped model that is shown again in Figure 4.17. This section uses the performed simulations to quantify the different parameters of this lumped model. The substrate resistances and the resistance of the ground interconnect are derived from the S-parameters provided by the EM simulation as explained in Chapter 3:

1. The S-parameters are first converted into Y-parameters.

2. Y_{ij} with $i \neq j$ corresponds to the admittance between port i and port j of the test structure. Hence, $\Re(1/Y_{ij})$ corresponds to the resistance between port i and port j.

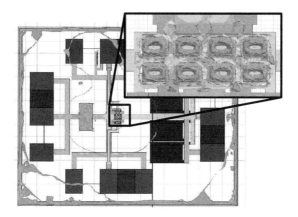

Figure 4.16 Simulated electric fields at 100 MHz. The darker regions correspond to the regions where the value of the electric field is high (i.e., where the P+ guard rings of the transistors are located). (See color section.)

The transistor parameters can be derived from the RF model of the transistor. The values of the parameters that determine the substrate noise coupling mechanism can be found in Figure 4.17.

Remember that the transfer function from the substrate contact to the drain of the transistor is given by:

$$TF_{sub2drain} = -\frac{R_{bulk2gnd}}{R_{sub2bulk}} \cdot g_{mb} \cdot R_{load} + \frac{R_{gnd}}{R_{sub2gnd}} \cdot g_m \cdot R_{load} \qquad (4.14)$$

In this case the load resistance (R_{load}) is the 50 Ω resistance of the measurement equipment. For this particular case this gives:

$$TF_{sub2drain} = -\frac{23\Omega}{1k\Omega} \cdot 34mS \cdot 50\Omega + \frac{R_{gnd}}{262\Omega} \cdot 317mS \cdot 50\Omega \qquad (4.15)$$

In the nominal case, the bulk effect is dominant because the resistance of the ground interconnect is only 0.1Ω. In that case, substrate noise currents couple at frequencies lower than 50 MHz into the bulk of the transistors and are drained toward the ground interconnect through the P$^+$ guard ring of the transistors (see Figure 4.14).

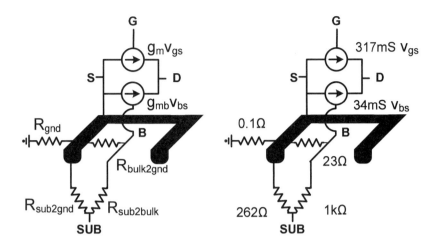

Figure 4.17 Lumped model of the test structure. S, D, B, and G reflect the source, drain, bulk, and gate terminals of the transistors. SUB indicates the substrate contact.

Increasing the resistance of the ground interconnect increases the importance of ground bounce when compared to the bulk effect because both effects have an opposite sign [see (4.14)]. There exists a value where both effects cancel each other out. Hence, the obtained isolation is optimal. The optimal value for the resistance of the ground interconnect can be calculated with (4.13). For this particular case the optimal value for R_{gnd} is approximately 0.65Ω. This means that the substrate noise isolation is worse for lower values of the resistance of the ground interconnect. This is very counterintuitive. One would expect that the isolation improves with decreasing values of the resistance of the ground interconnect. To demonstrate that this canceling effect really exists, a dedicated measurement experiment is set up that demonstrates that for increasing values of the resistance of the ground interconnect, the dominant coupling mechanism shifts from the bulk effect toward ground bounce.

4.5.5 Experimental Validation of the Substrate Noise Coupling Mechanisms

It is important to show with measurements that the two coupling mechanisms are present and that their importance depends on the resistance of the ground interconnect. In the nominal case, it is already demonstrated that the dominant coupling mechanism is the bulk effect, showing a phase shift of 180 degrees. In this case the impedance of the ground interconnect is very low: approximately 0.1Ω.

In order to demonstrate that for increasing values of the resistance of the ground interconnect the dominant coupling mechanism shifts from the bulk effect toward ground bounce, the resistance of the ground interconnect needs to be increased. It is not possible to change the resistance of the on-chip ground interconnect without reprocessing the test structure. To circumvent this problem a dedicated PCB is made. The PCB has two SMA connectors whose signal traces are shorted. A variable resistor is placed between their ground connections. Consequently the resistance of the ground interconnect can be increased from 0.5Ω up to 1.5Ω.

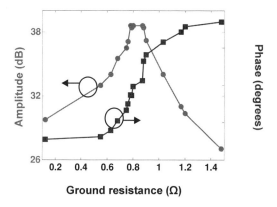

Figure 4.18 Measured isolation as a function of the resistance of the ground interconnect. The ● reflects the amplitude and the □ the phase.

Such a PCB is connected between each probe and the corresponding port of the measurement equipment. The S-parameters are measured for different values of the resistance of the ground interconnect. Figure 4.18 shows the measured isolation and phase at 100 kHz as a function of the resistance of the ground interconnect. This figure shows that for low values, where the bulk effect dominates, the isolation is 29 dB. The isolation starts to increase due to the increasing importance of the ground bounce effect. At a ground resistance of 0.8Ω, the isolation has an optimal value of 39 dB. This is an improvement of 10 dB. The corresponding phase is approximately 90 degrees. For ground resistances larger than 0.8Ω ground bounce starts to dominate. The isolation is decreasing. For a ground resistance of 1.4Ω the isolation is 27 dB. The corresponding phase shift is now close to zero. This corresponds to noise coupling into the source of the transistor.

This shift in coupling mechanism versus the ground resistance suggests that in some circuits the dominant coupling mechanism is the bulk effect while in other circuits ground bounce dominates. Millimeter-wave circuits that make use of a ground plane are more likely to suffer from the bulk effect. Circuit designs where the resistance of the ground interconnect is larger that 1Ω will suffer from ground bounce.

In Chapter 5, we will investigate the impact of substrate noise on analog/RF circuits. It will be shown that the dominant coupling mechanism in a 48–53 GHz LC-VCO is the bulk effect. This LC-VCO uses a ground plane. This ground plane provides a very low impedance to the PCB ground. The next chapter also demonstrates that the dominant coupling mechanisms in a 900 MHz LC-VCO is ground bounce. Here the value of the resistance of the ground interconnect is around 6Ω.

4.6 CONCLUSIONS

In this chapter we presented a powerful methodology to predict the impact of substrate noise on active devices. The proposed methodology requires a lot of information that is easily available to the designer:

- The layout of the active devices and the interconnects.

- The conductivity of the doped regions and the metal layers. When N-doped regions are considered, the different PN junction capacitances are required.

- The RF models of the transistors.

The impact of substrate noise on active devices is predicted by combining two different tools. In order to use this methodology, the designer needs:

- *An EM simulator:* The EM simulator models the substrate and the interconnects.

- *A circuit simulator:* The corresponding model of the EM simulation is elegantly included into one simulation model together with the RF models of the active devices. On this simulation model the analog designer can apply any circuit analysis.

The methodology is illustrated to the analog designer by the means of two examples: First, it is demonstrated that the methodology is able to characterize a single transistor up to 110 GHz. In the second example, the methodology is used the predict the different coupling mechanisms in a transistor. For this example it is shown that for values of the ground resistance lower than 0.65Ω substrate noise couples in to the bulk of the transistor. For values of the ground resistance higher than 0.65 Ω, the dominant coupling mechanism is ground bounce. As both effects have an opposite sign, the substrate noise isolation is optimal for a value of the ground interconnect of 0.65Ω. This value can be used as a guideline to determine whether the circuit will suffer more from the bulk effect than from ground bounce.

Now that the designer has the means to characterize transistors and to predict the impact of substrate noise on transistors, the complexity can be increased and the methodology can be used to predict the impact of substrate noise on analog/RF circuits.

References

[1] P. R. Gray, P. J. Hurst, S. H. Lewis, and R. Meyer, *Analysis and Design of Analog Integrated Circuits*, New York, John Wiley & Sons, 2001.

[2] Y. Zinzius, E. Lauwers, G. Gielen, and W. Sansen, "Evaluation of the substrate noise effect on analog circuits in mixed-signal designs," *Proc. SSMSD Southwest Symposium on Mixed-Signal Design 2000*, February 27–29, 2000, pp. 131–134.

[3] Substrate Noise Analysis Cadence, http://www.cadence.com.

[4] S. Ponnapalli, N. Verghese, W. K. Chu, and G. Coram, "Preventing a 'noisequake' [substrate noise analysis]," *IEEE Circuits and Devices Magazine*, Vol. 17, No. 6, 2001, pp. 19–28.

[5] HFSS, http://www.ansoft.com/products/hf/hfss/.

[6] Spectre RF, http://www.cadence.com/products/custom_ic/spectrerf.

[7] D. Lovelace, J. Costa, and N. Camilleri, "Extracting small-signal model parameters of silicon MOSFET transistors," *Proc. IEEE MTT-S International Microwave Symposium Digest*, May 23–27, 1994, pp. 865–868.

[8] R. Chang, M.-T. Yang, P. Ho, Y.-J. Wang, Y.-T. Chia, B.-K. Liew, C. Yue, and S. Wong, "Modeling and optimization of substrate resistance for RF-CMOS," *Electronic Devices*, Vol. 51, No. 3, 2004, pp. 421–426.

[9] R. Vinella, G. Van der Plas, C. Soens, M. Rizzi, and B. Castagnolo, "Substrate noise isolation experiments in a 0.18 μm 1P6M triple-well CMOS process on a lightly doped substrate," *Proc. IEEE Instrumentation and Measurement Technology*, 2007, pp. 1–6.

[10] D. White and M. Stowell, "Full-wave simulation of electromagnetic coupling effects in RF and mixed-signal ICs using a time-domain finite-element method," *IEEE Transactions on Microwave Theory and Techniques*, Vol. 52, No. 5, 2004, pp. 1404–1413.

Chapter 5

Measuring the Coupling Mechanisms in Analog/RF Circuits

5.1 INTRODUCTION

Analog/RF circuits are prone to substrate noise coupling. In general, a noise signal that couples into an analog/RF circuit will modulate with the desired analog/RF signal both in frequency and amplitude. Those modulation effects cause sideband spurs to appear around the desired analog/RF signal and its harmonics [1, 2]. Furthermore, the noise signal also couples directly to the output in a linear way without frequency translation by the analog/RF circuit. To illustrate that all those spurs are indeed present, a 900 MHz (f_{LO}) LC-VCO is considered when a 304 MHz sinusoidal is injected into the substrate (f_{sub}). Figure 5.1 shows the spectrum at the output of the 900 MHz VCO. Note the sideband spurs around the local oscillator ($f_{LO} \pm f_{sub}$) and its second harmonic ($f_{LO} - 2 \cdot f_{sub}$), and note also the direct coupled spur (f_{sub}) at the output of the VCO.

In general, the different spurs can be caused by two types of aggressors:

- Switching activity of the digital circuitry;

- The signal of an adjacent analog circuit such as the transmitted signal of a power amplifier.

By means of a real-life example of an analog/RF circuit that is perturbed by each of the aggressors mentioned above, the reader is shown how severe the

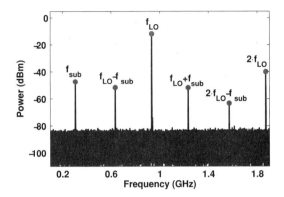

Figure 5.1 Measured spectrum of a 900 MHz LC-VCO when a sinusoidal signal with a frequency of 304 MHz is injected into the substrate.

analog/RF circuitry suffers from the presence of noise coupling. As an example, an LC-VCO is considered as a victim, which is perturbed by an aggressor.

In the first case, an LC-VCO is aggressed by the switching activity of some digital circuitry. Figure 5.2 shows the spectrum of a 3 GHz LC-VCO that is perturbed by the switching activity of a 40k gate digital modem [1]. Note the presence of several spurious tones close to the carrier.

In the second case, another LC-VCO is perturbed by a power amplifier. Both circuits are part of a large transceiver. Figure 5.3(a) shows the output spectrum of the 4.8 GHz LC-VCO when the power amplifier (PA) is turned off. Note the spectral purity of the oscillator. When the PA is turned on and excited with a 2.4 GHz WLAN signal, the spectral purity of the oscillator is ruined, see Figure 5.3(b).

In general, the type of spur that degrades most of the performance of the analog/RF circuit depends on the analog/RF circuit itself. If one considers a low-pass filter that is perturbed by the switching activity of the digital circuitry, the direct coupled spur is the most harmful to the filter because this spur may lie in the passband of the filter. In the case of a VCO, the sideband spurs degrade most the performance of the VCO [3]. This is why in the examples shown above, only the sideband spurs are shown. The sideband spurs close to the LO are especially harmful because they degrade the phase noise performance of the LC-VCO [2]. At higher frequencies the sideband spurs around the LO and second harmonic of the LO and also the direct coupled spurs can be very harmful. For example, FDD (frequency division duplex) transceivers are very prone to substrate noise coupling.

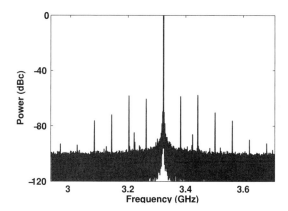

Figure 5.2   The switching activity of the digital 40k modem creates many unwanted spurs in the spectrum of the 3 GHz LC-VCO. The clock frequency of the modem is 20 MHz.

Such a transceiver transmits and receives simultaneously in different frequency bands. Brenna et al. [4] reported an FDD receiver where transmit and receiver band are separated by 130 MHz. High-frequency spurs of the VCO that are transmitted by the PA are located in the frequency band of interest of the receiver. Hence, high frequency spurs can cause the violation of spectral masks. When an analog/RF design is malfunctioning because of substrate noise coupling, the designer is often left without a clue of how to solve the problem.

This chapter exploits different measurement techniques to retrieve the dominant coupling mechanisms for two aggressor cases. In the first case, it is shown that the combination of different types of measurements enables one to extract a large amount of information that is necessary to interpret how substrate noise couples into a sensitive analog/RF circuit.

The second case considers an analog/RF that is aggressed by a power amplifier. The transmitted signal of the PA causes unwanted spurs to arise in the spectrum of the analog/RF circuit. Those spurs are caused by different coupling mechanisms. Dedicated experiments are performed to reveal the importance of the different coupling mechanisms between both RF circuits.

Both examples will determine what is required to model the noise coupling mechanisms in analog/RF circuits accurately. The modeling of the noise coupling mechanisms is the subject of Chapter 6.

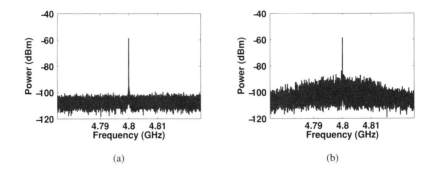

(a)   (b)

Figure 5.3 A 4.8 GHz LC-VCO is perturbed by the transmitted RF signal of the power amplifier. (a) Spectrum of a 4.8 GHz LC-VCO when the PA is turned off. (b) Spectrum of a 4.8 GHz LC-VCO when the PA is turned on.

5.2 MEASUREMENT-BASED IDENTIFICATION OF THE DOMINANT SUBSTRATE NOISE COUPLING MECHANISMS

An analog/RF circuit cannot be considered as a single noise reception node. The analog circuitry has many reception nodes where substrate noise can couple into. Pinpointing the dominant noise coupling mechanism in an analog/RF circuit based on measurements is not a trivial task. The major difficulty is that only the signals that are present at the connections of the PCB can be sensed and thus the reception nodes themselves cannot be measured directly. At first sight, it seems to be impossible to identify the on-chip substrate noise phenomena based on the *remote* measurements. However, combining different types of measurements enables the designer to extract a large amount of information that is necessary to identify how the noise signal propagates and how it couples into the sensitive analog/RF circuit [5].

The previous section showed that when a substrate noise couples into the analog/RF circuit, different spurs arise. First, different measurement techniques are proposed to measure the different spurs and distinguish whether the sideband spurs are modulated in amplitude or in frequency. Second, sensitivity functions are introduced. Sensitivity functions determine how the spurs are influenced when a perturbation is applied on a certain reception node of the circuit [6]. Using sensitivity functions enables one to determine the dominant coupling mechanisms. Finally, it is shown how to determine the influence of the PCB on the substrate noise coupling mechanisms.

In the next section, those measurement techniques are demonstrated on a real-life silicon example.

5.2.1 Measurement of the Different Spurs

The aim of these measurements is to measure the transfer characteristic between the input signal (entering the substrate) and the output spectrum of the analog/RF circuit. The amplitude information of the frequency response provides information about the substrate noise behavior and about the type of modulation of the sideband spurs. Measurements with a high dynamic range are required due to the relatively high attenuation of the transfer characteristic. This can be seen in Figure 5.1, which clearly shows that in the special case of a VCO a dynamic range of at least 60 dB is required to accurately measure the spurious tones. One will now consider two cases depending on the frequency dependence between excitation and response. In the first analysis, the frequency of the perturbation is equal to the frequency difference between the analog/RF signal frequency and the response. In general, those frequencies are different. The output frequency, labeled f_{out}, is defined by $f_{out} = k f_{RF} \pm f_{sub}$, with f_{RF} the frequency of the analog/RF signal, f_{sub} the frequency of the exciting substrate signal, and k being zero or not.

The *amplitude transfer characteristics of all spurs* are measured using a spectrum analyzer while the frequency of the substrate noise signal is varied. The main advantage of the spectrum analyzer is its large available dynamic range. The main disadvantage is the lack of phase information for the measured tones.

The *transfer function of the direct coupling* ($k = 0$) can easily be measured using classical vectorial network analyzers. These measurements provide both amplitude and phase information at a high dynamic range.

The *phase information of the modulated spurs* ($k \neq 0$) cannot be determined by a linear network analyzer or a spectrum analyzer. Therefore, the type of modulation can be revealed by feeding the output of the VCO to a limiting amplifier and measuring the output of the limiter with a spectrum analyzer. Such a limiter removes the spurs resulting from AM because the modulated signal does not have a constant envelope. However, the limiter will not remove the spurs resulting from FM because the modulated signal has a constant envelope. The spurious tones are caused by variations of the zero crossings at the output of the analog/RF circuit. When such a limiting amplifier is not available, one can use a sampling oscilloscope to reveal the type of modulation. The output of the analog/RF circuit is measured with a sampling oscilloscope. In a first step, the measured output is Fourier transformed using MATLAB [7]. In a second step, the measured output is first limited using MATLAB and then Fourier transformed. Then, both spectra are compared with each other. The spurs that are suppressed by the limiting operation are modulated in amplitude.

5.2.2 Sensitivity Functions

In order to reveal the dominant coupling mechanisms in a certain frequency region, sensitivity functions are introduced. Sensitivity functions determine how circuit aspects like the gain, oscillation frequency, and cutoff frequency vary when a perturbation is applied on a substrate reception node of the circuit. Sensitivity functions can both describe the linear transfer characteristic from the reception node to the output of the analog/RF circuit (direct coupled spurs) as well as the transfer characteristic of periodically time-varying circuits (sideband spurs). As the substrate perturbation can safely be assumed to be small, this dependence can be linearized:

$$\Gamma(V_{i\phi} + \Delta V_i) = \Gamma(V_{i\phi}, V_{tune}) + \left.\frac{\partial \Gamma}{\partial V_i}\right|_{(V_{i\phi})} \cdot \Delta V_i \qquad (5.1)$$

$$\Gamma(V_{i\phi} + \Delta V_i) = \Gamma(V_{i\phi}, V_{tune}) + K_i \cdot \Delta V_i \qquad (5.2)$$

The sensitivity functions are defined by sensitivity factors K_i: the sensitivity factor K_i related to a reception point in the analog/RF circuit i describes the variation of the circuit aspect Γ due to a variation of the voltage V_i on this reception point i with a nominal value $V_{i\phi}$. The chosen circuit aspect Γ depends on the analog/RF circuit that is considered. Remember that only the external voltages can be varied during measurements. In order to demonstrate how to choose the sensitivity factors K_i, a few examples are given:

- The LNA is a sensitive RF circuit since every signal that couples into the LNA is amplified through the whole receiver chain. Hence, the sensitivity factor of an LNA due to a variation on the supply voltage (V_{DD}) can be defined as the variation of the LNA gain (G) due to a voltage variation ΔV_{DD} on a reception point V_{DD}:

$$K_{VDD} = \left.\frac{\partial G}{\partial V_{VDD}}\right|_{V_{VDD\phi}} \qquad (5.3)$$

where $V_{DD\phi}$ is the nominal value of the power supply.

- The sensitivity factor of a lowpass filter due to a variation on the ground voltage V_{GND} can be defined as the variation of the cutoff frequency (ω_c) due to a voltage variation V_{GND} on a reception point GND:

$$K_{GND} = \left. \frac{\partial \omega_c}{\partial V_{GND}} \right|_{V_{GND\phi}} \tag{5.4}$$

- The sensitivity factor of a VCO can be defined as the variation of the LO-frequency (f_{LO}) due to a voltage variation V_i on a reception point i.

In general, the circuit aspect Γ of an analog/RF circuit to a substrate noise perturbation, is determined by every voltage V_i on any entry point i that changes the value of Γ. V_i can be expressed as follows:

$$V_i(t) = \int_0^t (h^i_{sub} * V_{sub})(t) d\tau \tag{5.5}$$

Here is h^i_{sub} the transfer function from the point where substrate noise originates (sub) to the reception point i of the analog/RF circuit. $V_{sub}(t)$ is the signal generated by the digital circuitry. For simplicity reasons, the substrate noise signal consists of a sine wave that has an amplitude A_{sub} and frequency f_{sub}:

$$V_{sub}(t) = A_{sub} \cdot \cos(2\pi f_{sub} t) \tag{5.6}$$

The circuit aspect Γ of the analog/RF circuit subject to a substrate noise signal (ΔV_{sub}) can be expressed as follows (see Figure 5.4):

$$\Gamma(V_{sub}) = \Gamma(V_{sub,\phi}) + \sum_i^n A_{sub} \cdot H^i_{sub}(f_{sub}) \cdot K_i \tag{5.7}$$

From this equation we see that the importance of the contribution of a sensitive node i to the overall impact thus depends on the product of both the amplitude of the noise signal (A_{sub}), the attenuation by the substrate $H^i_{sub}(f)$, and the sensitivity function K_i. Hence, in general, there are three ways to reduce the substrate noise coupling:

1. Reducing the amplitude of the noise signal. There exists in literature different ways to reduce the generation of substrate noise [8, 9].

2. Attenuating the propagation of substrate noise. Chapter 3 proposes different types of guard rings, which isolate the analog circuitry from its digital aggressor.

3. Making the analog circuitry more immune to the impact of substrate noise.

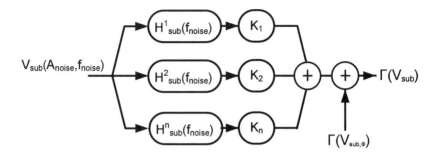

Figure 5.4 The degradation of a circuit aspect depends on the substrate noise amplitude, the substrate transfer function, and the sensitivity of the analog circuitry to substrate noise.

The immunity of the analog circuitry to substrate noise can only be effectively increased if the dominant coupling mechanisms are revealed. Therefore as much information as possible about the substrate noise coupling mechanisms needs to be gathered.

5.2.3   Determining the Influence of the PCB

The IC consisting of both the analog and digital circuitry are usually mounted on a PCB. Besides the IC itself, other components are also mounted on this circuit. The behavior of these additional PCB components is frequency dependent and hence they can influence the substrate noise behavior. Fortunately, at the PCB level, components can easily be removed or replaced. To determine if the PCB components influence the substrate noise behavior, we propose to remove certain PCB components or to replace them with other PCB components. After this PCB modification, the substrate noise spurs need to be remeasured. Comparing the spurs before and after the PCB modification allows one to determine if they influence the substrate noise behavior. For example, PCB decoupling capacitors can easily be removed or changed by a decoupling capacitor with a different value. In this way, one can check if the PCB decoupling capacitors change the substrate noise behavior of the IC. If a certain PCB component influences the substrate noise behavior, the component itself can be measured. Those measurements can provide more information about how the substrate noise behavior is influenced by the PCB components.

5.3   EXAMPLE: 900 MHZ LC-VCO

A VCO has been chosen as an example since it is a very sensitive analog/RF circuit and is therefore an ideal victim to study the noise sensitivity. In this study, the focus is put on the analysis of the substrate noise coupling mechanisms in the case where the VCO is perturbed by the digital switching activity. First, the design of the VCO is discussed. There are two reasons for this: an in-depth understanding of the circuit is required to understand how a signal propagates through the circuit. Also, we want to show that the circuit has a good performance and is therefore a good candidate to study the impact of substrate noise. Then the three measurement techniques mentioned above are applied to this example:

- Measurement of the different spurs;

- Using the sensitivity functions to determine the dominant coupling mechanisms;

- Determining if and how the PCB components influence the substrate noise behavior.

5.3.1   Description of the LC-VCO

In order to understand the different substrate noise coupling mechanisms, it is mandatory that the designer understands how the analog/RF circuit works. Therefore the analog/RF circuit under test is briefly described. The analog/RF circuit under test is a 900 MHz LC-tank VCO, designed in a 0.18 μm CMOS technology on a lightly doped substrate with 20 Ωcm resistivity. It consists of a cross-coupled NMOS-PMOS transistor pair and a PMOS current mirror (see Figure 5.5). This topology is popular for its low phase noise potential due to the lower flicker noise of the PMOS devices when compared to NMOS and for its reduced sensitivity to power supply noise. The VCO uses a fully integrated inductor with an inductance of 16 nH and a Q factor of 9. The VCO core draws 1.2 mA supply current from a 1.8 V power supply. A worst case phase noise of -90 dBc/Hz at 100 kHz offset is achieved in this design. This is acceptable given the low inductor Q factor and the rather low current consumption. The accumulation mode NMOS varactors (C_{min} = 1.5 pF and C_{max} = 4.5 pF) are used to tune the VCO frequency from 750 MHz to 1.05 GHz. The VCO core is buffered by inverters. Bypass capacitors are added at the output to decouple the DC. The output power level of the local oscillator is -12 dBm.

The VCO is mounted on a PCB. External connections are foreseen on the PCB to measure the output of the VCO and to inject a test signal into the substrate

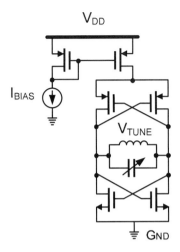

Figure 5.5   Schematic of the 900 MHz LC-VCO.

through a dedicated on-chip substrate contact. An RF perturbation signal of 10 dBm is then injected in the substrate. This seemingly high power level has been selected such that a sufficient signal to noise ratio (SNR) is obtained for all the experiments but is still sufficiently low to maintain a linear behavior of the VCO in response to the perturbing signal. Note that the reflection coefficient of the substrate contact is as high as -2.2 dB at 100 MHz. To avoid excessive reflections in the cables of the DC lines, bias tees are used to provide an RF termination of 50 Ω.

The supply lines are optionally decoupled using 100 pF, 100 nF, and 100 μF off-chip SMD decoupling capacitors to investigate the impact of the off-chip decoupling. In addition, a fixed 19 pF decoupling capacitor was foreseen on-chip.

5.3.2   Substrate Sensitivity Measurements

The *amplitude transfer function of all spurs* is measured using a spectrum analyzer when a 10 dBm signal with frequency range from 100 MHz to 900 MHz is injected into the substrate. Those measurements are performed first with and then without the off-chip decoupling capacitors. Both the responses of the LC-VCO are shown in Figures 5.6 and 5.7. The figures include the direct coupling (f_{sub}), the modulated spurs around the first harmonic of the VCO ($f_{LO} \pm f_{sub}$), and the left sideband around the second harmonic of the VCO ($2f_{LO}\text{-}f_{sub}$). These measurements clearly demonstrate the dependency of the off-chip decoupling on the transfer characteristic. One can also observe that at offset frequencies lower than 150 MHz, the sideband spurs are increasing at a very high rate. In that frequency region, the substrate noise signal starts to pull the local oscillator of the VCO. When the oscillator is perturbed by a substrate noise signal close to the LO frequency, the LO frequency becomes identical to that of the perturbing signal. The VCO locks to the perturbing signal.

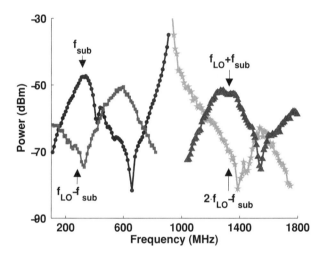

Figure 5.6   Measured output powers of the LC-VCO with 100 μF, 100 nF, and 100 pF off-chip decoupling capacitors.

The impact of the decoupling capacitors is determined by measuring the direct coupling for various decoupling configurations, namely:

- Without off-chip decoupling;

- 100 μF in parallel with a 100 nF decoupling capacitor;

- A set of 100 μF, 100 nF, and 100 pF capacitors.

The various measurements shown in Figure 5.8 clearly indicate an influence of the off-chip decoupling onto the substrate sensitivity of the LC-VCO. In spite of what is intuitively expected, adding 100 pF in parallel with 100 μF and 100 nF does not improve the immunity against substrate noise. Also note that the common practice

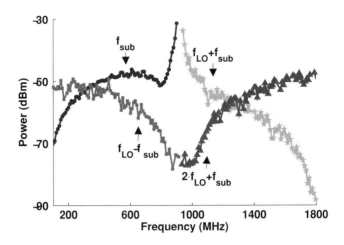

Figure 5.7   Measured output powers of the LC-VCO without off-chip decoupling capacitors.

of using a large decoupling capacitor in parallel with a smaller one (e.g., 100 μF in parallel with 100 nF) does not lead to a significant reduction of the substrate noise sensitivity either. This will be explained in Section 5.3.5.

To understand the impact mechanism of the substrate perturbation, a better knowledge is required of the type of modulation which occurs in the VCO. Figure 5.9 shows that the spurs left and right of the LO have the same amplitude below 10 MHz. This implies that both spurs result from a single type of modulation, as one expects from pure AM and FM. The modulation type changes at higher frequencies (10 MHz–100 MHz) as can be seen in the spectrum analyzer measurements. The measurements between 10 MHz and 100 MHz reveal that the modulation is no longer symmetrical around the carrier. Above 100 MHz the spurs are again symmetrical around the LO. The type of modulation can be revealed by feeding the output of the VCO to a limiting amplifier. Such a limiting amplifier removes the AM modulated spurs. Figure 5.10 shows that the impact mechanism shifts from FM toward AM. Starting from 10 MHz the amplitude of the spurs after limiting is smaller than the amplitude of the spurs without limiting. Above 100 MHz the dominant impact mechanism is AM. The next sections reveal the dominant coupling mechanisms for both the FM and the AM modulated spurs. Section 5.3.3 introduces FM sensitivity functions. Studying those FM sensitivity functions will reveal the dominant coupling mechanism for the FM modulated spurs. The nature of the AM

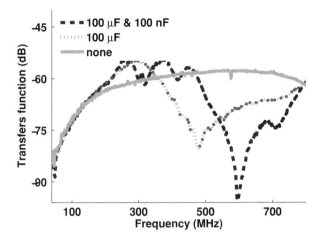

Figure 5.8   Direct coupling for various decoupling strategies.

modulated spurs is revealed in Section 5.3.4. This section formulates a hypothesis about the dominant AM coupling mechanism. This hypothesis is verified afterwards with measurements.

5.3.3   Revealing the Dominant Coupling Mechanism for FM Spurs

In order to reveal the dominant coupling mechanism in the frequency region where the spurs are FM modulated, FM sensitivity functions are introduced. The FM sensitivity function determines how much frequency modulation results from a perturbation that is applied on a certain node of the circuit. The sensitivity factor K_i is defined in the case of the VCO as the variation of the oscillator frequency f_{LO} due to a variation of the voltage V_i on this entry point i with a nominal value $V_{i\phi}$.

$$K_i(V_{tune}) = \left. \frac{\partial f_{LO}(V_{tune})}{\partial V_i} \right|_{V_{i\phi}} \tag{5.8}$$

As an example, the sensitivity of the VCO to ground disturbances is given by the K_{GND} of the VCO:

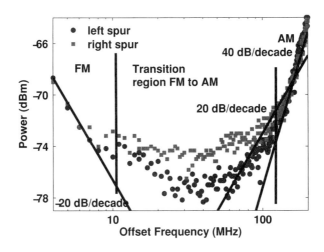

Figure 5.9   The measured left and right spurs ($k = -1$ and $k = 1$) clearly show the transition region between 10 and 100 MHz.

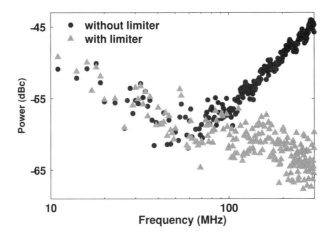

Figure 5.10   Limiter experiment which clearly illustrates that the modulation type is changing from FM toward AM modulation.

$$K_{GND}(V_{tune}) = \left. \frac{\partial f_{LO}(V_{tune})}{\partial V_{GND}} \right|_{V_{GND\phi}} \tag{5.9}$$

V_{GND} is the voltage on the ground interconnect. $V_{GND\phi}$ is the unperturbed version of this voltage. Similar expressions can be found for the sensitivity with respect to the other external nodes of the circuit such as V_{DD}, V_{tune}, and V_{bias} of the VCO.

Those sensitivity functions can easily be measured when a small voltage difference at the different terminals of the VCO is applied and observing the shift in oscillation frequency [see (5.10)]. This measurement is then repeated for different values of the tuning voltage. Hence, (5.9) is approximated by a difference equation:

$$K_{V_{DD}}(V_{tune}) = \frac{\Delta f_{LO}(V_{tune})}{\Delta V_{V_{DD}}(V_{tune})} \tag{5.10}$$

The measurement of the FM sensitivity function of the ground K_{GND} is more complex as it has to be performed indirectly. The VCO ground is connected to the PCB ground with a bonding wire, which is considered as the absolute reference, and no voltage source can be placed in between both. In the absence of a substrate perturbation, the voltage on four nodes can modulate the oscillation frequency: V_{tune}, V_{DD}, V_{bias}, and V_{GND}. In this case:

$$\Delta f_{LO} = K_{GND} \cdot \Delta V_{GND} + \Delta K_{V_{DD}} \cdot \Delta V_{V_{DD}}$$
$$+ K_{V_{tune}} \cdot \Delta V_{V_{tune}} + K_{bias} \cdot \Delta V_{bias} \tag{5.11}$$

If $\Delta V_{tune} = \Delta V_{DD} = \Delta V_{bias} = \Delta V_{GND} = \Delta V$, the voltages present on the VCO will remain perfectly constant despite the applied changes because the behavior of the circuit depends on the applied voltage difference. Thus, if one applies the same voltage difference ΔV at all the terminals of the VCO, the behavior of the VCO remains unchanged and no modulation of the local oscillation frequency will occur, and:

$$\Delta V \cdot (K_{GND} + K_{V_{bias}} + K_{V_{DD}} + K_{V_{tune}}) = 0 \tag{5.12}$$

The sensitivity of the ground terminal K_{GND}, which reflects the sensitivity to ground bounce can be simply determined by summing the measured sensitivities of the V_{tune}, V_{DD}, and V_{bias} terminal:

$$K_{GND} = -K_{V_{bias}} - K_{V_{DD}} - K_{V_{tune}} \tag{5.13}$$

Figure 5.11 shows the sensitivity function of the different terminals of the VCO under test. This figure indicates that the V_{DD} terminal of the VCO is less sensitive than the other terminals.

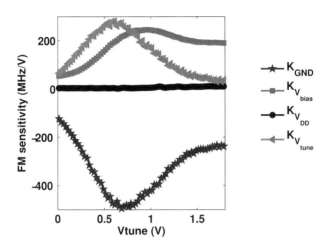

Figure 5.11   The FM sensitivities of the different terminals of the 900 MHz LC-VCO.

In order to calculate the power of the FM-induced spurs, we consider a sinusoidal substrate noise signal with an amplitude A_{sub} and frequency V_{sub}:

$$V_{sub}(t) = A_{sub} \cdot \cos(2\pi f_{sub} t) \tag{5.14}$$

The output voltage $V_{out}(t)$ of the VCO subject to FM can be expressed as follows:

$$V_{out}(t) = A_{LO} \cdot \cos\left(2\pi f_{LO}t + 2\pi \sum_{i}^{n} K_i(V_{tune}) \int_0^t (h_{sub}^i * V_{sub})(t)dt\right) \tag{5.15}$$

Here $A_{LO}(V)$ represents the local oscillator amplitude and K_i (Hz/V), defined in (5.9), and the sensitivity of the local oscillator frequency to a voltage variation V_i, defined in (5.5). N is the number of entry points through which substrate noise can enter the VCO. h_{sub}^i is the transfer function from the point where substrate noise originates, to the reception point i in the VCO. Since the substrate noise

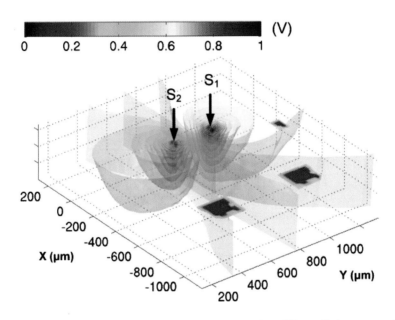

Figure 2.7 Equipotential surfaces at DC in the substrate when 1V is applied to contact S_1 by a 50Ω terminated DC source. Contact S_2 is terminated with a 50Ω resistance.

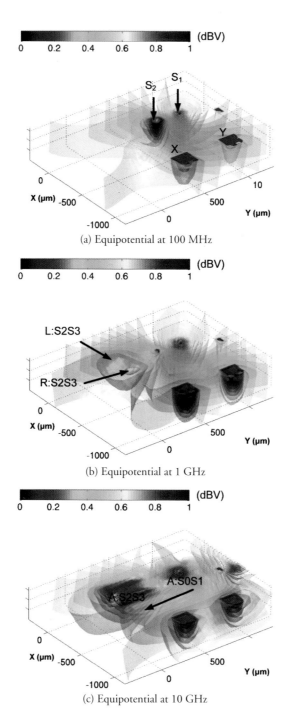

Figure 2.9 (a–c) 3-D equipotential surfaces when 1V is applied to contact S_1 and 0V is applied to contact S_2. Note that the unit is dBV.

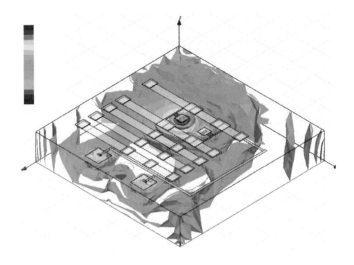

Figure 2.16 Simulated electrical field distribution at 100 MHz. The electrical field distribution provides the same insight in the substrate noise propagation as the equipotential surfaces do.

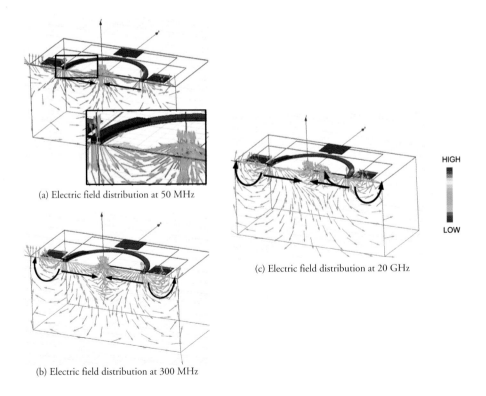

(a) Electric field distribution at 50 MHz

(c) Electric field distribution at 20 GHz

(b) Electric field distribution at 300 MHz

Figure 3.5 (a–c) Electric field distribution at different frequencies.

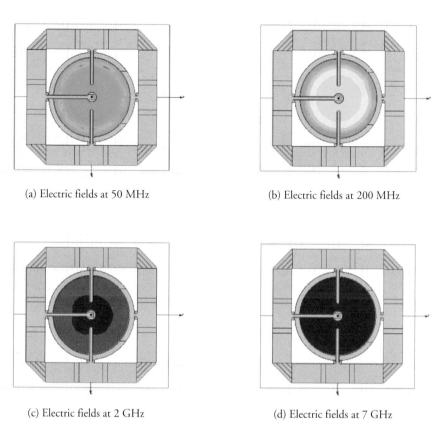

(a) Electric fields at 50 MHz

(b) Electric fields at 200 MHz

(c) Electric fields at 2 GHz

(d) Electric fields at 7 GHz

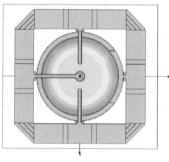

(e) Electric fields at 20 GHz

Figure 3.10 The electric field distribution clearly visualizes the capacitive coupling through the n-well. The red regions correspond to high values of the electric field, and the blue regions correspond to low values of the electric field. The pseudo-coloring reflects a logarithmic scale.

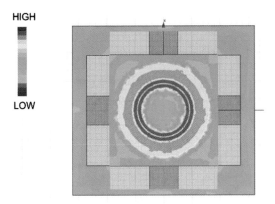

Figure 3.13 Electrical field distribution in the p-well of the P⁺ guard ring isolation structure at a frequency of 50 MHz. The red regions correspond to high electrical field values. Blue regions correspond to low electrical field values. The pseudo-coloring reflects a logarithmatic scale.

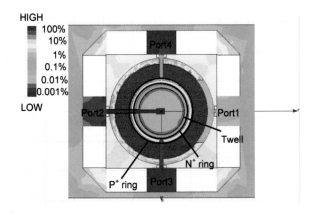

Figure 3.17 Electrical field distribution at 50 MHz.

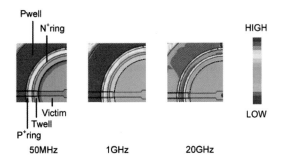

Figure 3.18 Electrical field distribution at 50 MHz, 1 GHz, and 20 GHz.

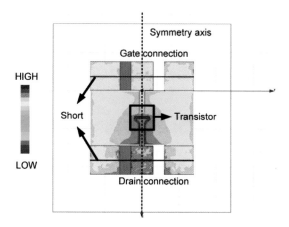

Figure 4.7 The electrical field distribution present in the transistor test bench. The field distribution is symmetrical.

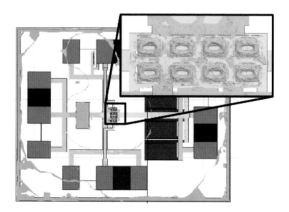

Figure 4.16 Simulated electric fields at 100 MHz. The darker regions correspond to the regions where the value of the electric field is high (i.e., where the P^+ guard rings of the transistors are located).

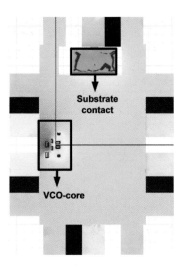

Figure 6.14 Simulated electrical fields at 100 MHz.

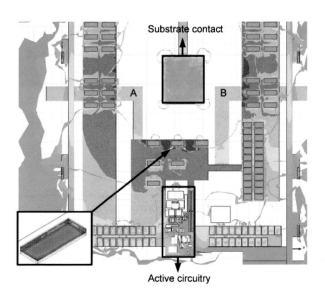

Figure 7.13 Simulated electric fields in the ground interconnect, the p-wells, and the n-wells.

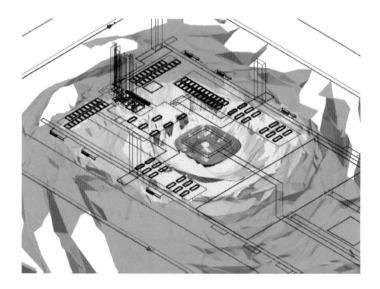

Figure 7.14 The current flow in the substrate is multidimensional.

signals are small compared to the local oscillator signal, narrowband FM can be assumed and the frequency domain expression for the spurious tones in the output spectrum of the VCO becomes (see Appendix A):

$$|V_{out}(f_{LO} \pm f_{noise})| = |\sum_{i}^{n} H^i_{sub}(f_{noise})K_i(V_{tune})\frac{A_{LO} \cdot A_{noise}}{2f_{noise}}| \quad (5.16)$$

From this equation we see that the importance of the contribution of a sensitive node or entry point i, for example the V_{DD} terminal, to the overall impact depends on the product of the attenuation by the substrate $H^i_{sub}(f)$ and the sensitivity function $K_i(V_{tune})$. At very low frequencies (< 1 MHz) it is assumed that $H^i_{sub}(f)$ is frequency independent. This means that substrate noise couples resistively into the LC-VCO. Capacitive coupling is negligible for frequencies lower than 1 MHz since the V_{DD} contact, the PMOS transistors, and the varactors are located in small N wells that have a small capacitance to the substrate. The coupling to the inductors in the tank is largely reduced by poly shielding.

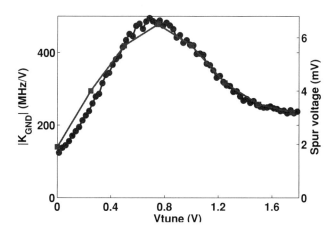

Figure 5.12   The spur voltage, marked with □ is at a frequency of 1 MHz proportional with the K_{GND} of the VCO, marked with ●.

In order to reveal the dominant substrate noise entry point with measurements, the sensitivity functions $K_i(V_{tune})$ for all the entry points are compared with the power of the spurs measured at 1 MHz offset frequency (see Figure 5.12). As can be seen in Figure 5.12 the voltage of the spurs is proportional with the sensitivity function K_{GND}. Therefore it can be concluded that ground bounce is the dominant substrate noise coupling mechanism and (5.16) can be simplified. Equation (5.17) can be used to calculate the power of the FM modulated spurs.

$$|V_{out}(f_{LO} \pm f_{noise})| = |H_{sub}^i K_{GND}(V_{tune}) \frac{A_{LO} \cdot A_{noise}}{2f_{noise}}| \qquad (5.17)$$

5.3.4 Revealing the Dominant Coupling Mechanism for AM Spurs

One can define AM sensitivity functions in order to reveal the dominant AM coupling mechanism. Those sensitivity functions can be defined in a similar way as the FM sensitivity functions:

$$K_i^{AM}(V_{tune}) = \left.\frac{\partial A_{LO}(V_{tune})}{\partial V_i}\right|_{(V_{i\phi})} \qquad (5.18)$$

However, it is very difficult to measure the very small amplitude variations on the LO signal. Therefore AM sensitivity functions are not used to reveal the dominant AM coupling mechanism.

To reveal the dominant AM coupling mechanisms, the behavior of the spurs is observed instead. A hypothesis is formulated to select the dominant AM coupling mechanism. Then a dedicated experiment is set up to verify the hypothesis.

Starting from 50 MHz the power of the spurs increases at a rate of 20 dB/decade (see Figure 5.9). At high frequencies (> 100 MHz) the power of the spurs is increasing at a rate of 40 dB/decade. In order to reveal why the spurs are increasing at such a high rate, the reader should note that:

- In a frequency independent system where the signals are modulated in amplitude, the power of the spurs does not vary with the offset frequency. Remember that this is not the case when the spurs are modulated in frequency. The power of FM spurs decreases by 20 dB/decade.

- The substrate behaves resistively up to 7.5 GHz [10], and hence the substrate needs to be considered as frequency independent at the studied frequencies.

Hence, the coupling mechanisms that cause an increase of the spurs at a rate of 20 dB/decade and later on 40 dB/decade are incorporated in the VCO itself. The hypothesis made is that substrate noise couples resistively via R_{sub} to the shield of the inductor (see Figure 5.14). A part of the coupled substrate currents is drained toward PCB ground through the ground connection of the shield of the inductor. The impedance of the ground connection is determined by the resistance and inductance of the on-chip tracks and the bond wire (R_{bond} and L_{bond}). At frequencies below 100 MHz, the impedance of the ground connection is dominated by the resistance R_{bond}. Another part of the substrate currents couples capacitively into the large inductor of the LC-tank via C_{ind}, which is the parasitic capacitance of the inductor. Figure 5.13 shows that this inductor occupies a large area of the chip. This explains the increase of the AM spurs in the frequency region between 10 MHz and 100 MHz at a rate of 20 dB/decade.

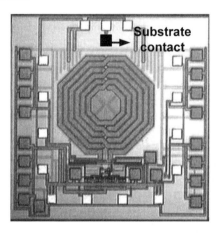

Figure 5.13   A large part of the 900 MHz LC-VCO is occupied by the on-chip inductor.

Above 100 MHz the impedance of the ground connection of the shield of the inductor is determined by the inductance of the bonding wire L_{bond}. With increasing noise frequencies, less substrate noise is drained toward the PCB ground and more substrate noise couples into the inductor of the VCO. The inductance of the bond wire (L_{bond}) and the parasitic capacitance of the inductor of the LC tank (C_{ind}) explains the increase at a rate of 40 dB/decade of the AM spurs for noise frequencies larger than 100 MHz.

It is very difficult to verify the hypothesis with measurements. However, a dedicated experiment has been set up to show that above 100 MHz power of the substrate noise induced spurs increases with the impedance of the bonding wire connected to the shield. In the VCO design, the shield of the inductor is connected with a separate bonding wire to the PCB ground. Different PCBs with different contact resistances of the bonding wires have been measured in order to show the

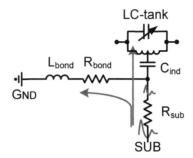

Figure 5.14   Substrate noise can couple capacitively to the inductor of the LC-VCO.

influence of this resistance on the impact of substrate noise. The resistances of the
bonding wires have been measured separately with DC probes. DC probes have
been placed on unused ground bonding pads located close to the bonded ground
pad. A DC current is then injected into the nonideal ground. It is assumed that most
of the current flows through the bonding wire. This is true since the impedance of
the bonding wire is at least one order of magnitude smaller than the impedance
of the other current paths. The measured DC resistance of one PCB is 6.6 times
larger than for the other PCB. The transfer function from the injected substrate
noise signal to the direct coupled spur at the output of the VCO is measured with
a network analyzer. The injected power is 10 dBm. Measurements with different
injected power levels show that the transfer function is independent of the injected
power level. The measurements are calibrated up to the connectors of the PCB.
Figure 5.15 shows that a PCB with a larger DC resistance of the bonding wire shows
a higher impact of 6 dB average. One can now safely assume that the proposed
mechanism is indeed the dominant one.

 Chapter 6 validates the assumption that substrate noise couples capacitively
into the inductor with simulations.

5.3.5   Influence of the PCB Decoupling Capacitors on the Substrate Noise Impact

Figure 5.8 clearly demonstrates the presence of a significant influence of the PCB
decoupling capacitors on the substrate noise impact. Adding the 100 μF decoupling
capacitors creates a parallel resonance in the transfer function at a frequency of
480 MHz. This resonance clearly improves the substrate noise immunity of the
VCO around this resonance frequency. However, at frequencies around 300 MHz

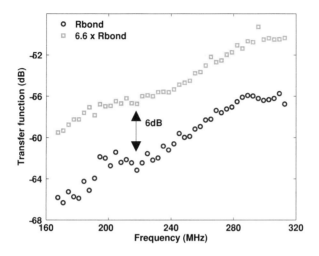

Figure 5.15 The impedance of the bonding wire plays an important role in the AM coupling mechanism.

the decoupling capacitors worsen the substrate noise immunity significantly. In order to better understand the mechanism that causes those resonances, the different decoupling capacitors are measured with an impedance analyzer.

Figure 5.16 shows the admittance of the capacitors from 1 MHz up to 1 GHz. Those measurements reveal that the 100 μF decoupling capacitors already behaves as an inductor for frequencies as low as 1 MHz. The 100 nF, and 100 pF decoupling capacitors have their self-resonance frequencies at, respectively, 35 MHz and 700 MHz. Above these frequencies, the components start to behave as an inductor. The DC lines are also decoupled with on-chip MIM capacitors. Those capacitors have a self-resonance frequency that is higher than 5 GHz. The measured resonance at a frequency of 480 MHz is the result of an LC-resonance of the parasitic inductance of the PCB decoupling capacitor, the PCB trace, and the bond wire with the 19 pF on-chip decoupling capacitors (see Figure 5.17). The frequency of the resonance is given by:

$$f_{res} = \frac{1}{2\pi \cdot \sqrt{(2 \cdot L_{bond} + L_{traces} + L_{100\mu F}) \cdot C_{on-chip}}} \qquad (5.19)$$

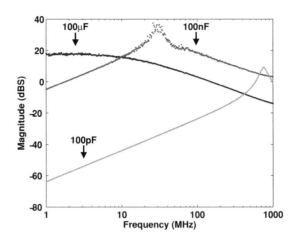

Figure 5.16   The measured admittance function from 1 MHz up to 1 GHz for a 100 pF, 100 nF, and a 100 μF SMD decoupling capacitor.

Adding the inductance values for the 100 nF decoupling capacitors, together with the 100 μF capacitors elevates the resonance frequency up to 580 MHz (see Figure 5.8). This is because the total parasitic inductance is lowered due to the parallel circuit connection of the parasitic inductance of the 100 μF and 100 nF decoupling capacitors and their traces; see (5.19). The 100 pF decoupling capacitors have almost no influence on the substrate noise impact. The impedance of the 100 nF decoupling capacitor is lower than the impedance of 100 pF in the frequency region up to 700 MHz.

5.3.6   Conclusions

In this section we have demonstrated that different types of measurements can be combined to extract the large amount of information that is necessary to interpret substrate noise coupling in analog/RF circuits.

The different measurement techniques have been used here to reveal the coupling mechanisms in the case of an LC-VCO. Substrate noise coupling in an LC-VCO causes spurs to appear. These spurs are modulated both in frequency and in amplitude. Measurements reveal that when noise is injected at low frequencies (<10 MHz) substrate noise coupling results in dominantly FM modulated spurs. At intermediate noise frequencies (between 10 MHz and 100 MHz) the type of

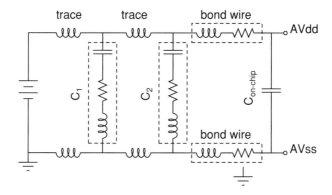

Figure 5.17 Circuit model for the decoupling and the bond wires when considering two decoupling capacitors C_1 and C_2.

modulation changes toward AM modulated spurs. The spurs are AM modulated at high frequencies (>100 MHz). At those frequencies the behavior of the substrate noise coupling is influenced by the parasitics of the PCB devices. Measurements reveal that the parasitics of the decoupling capacitors resonates with the on-chip decoupling capacitors.

Section 5.4 studies the case where the LC-VCO is aggressed by a power amplifier. The transmitted signal of a power amplifier causes unwanted spurs to appear in the spectrum of the VCO. Besides substrate noise coupling, other coupling mechanisms are responsible for those unwanted spurs. The next section will quantify the different coupling mechanisms and reveal their importance with measurements.

5.4 STUDY OF THE COUPLING MECHANISMS BETWEEN A POWER AMPLIFIER AND AN LC-VCO

RF transmitters are designed to convert baseband signals into RF signals using efficient modulation techniques. A VCO is thereby used as a local oscillator (LO) for the upconversion. This modulated RF signal is then amplified and transmitted by a power amplifier (PA). To maintain the high bandwidth of such transmitters it is mandatory to minimize the length of the interconnects. Hence, the VCO and the PA must be placed close to each other on the same die (see Figure 5.18).

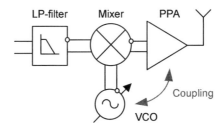

Figure 5.18 To maintain the high bandwidth of the transmitter, the VCO and the PA must be placed close to each other. This results in a potential high coupling of the PA with the sensitive VCO.

In such transmitters the PA acts as an aggressor while the VCO acts as a victim. The transmitted RF signal couples into the VCO and causes unwanted re-modulation of the VCO and subsequent spectral spreading of the desired RF signal. There are different possible coupling mechanisms that are responsible for possible failure of the RFIC. Theoretically, the PA can couple resistively, magnetically, and capacitively with the VCO. The possible existence of the different coupling mechanisms is determined next.

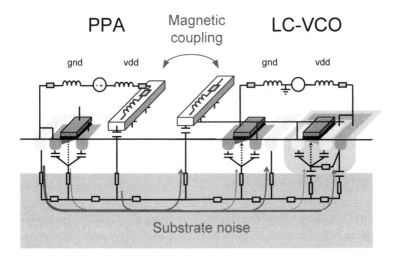

Figure 5.19 Different coupling mechanisms between a PPA and an LC-VCO.

- *Resistive coupling*: The PA and the VCO share the same die. Therefore they can couple resistively with each other through the common substrate (see Figure 5.19).

- *Magnetic coupling*: The PA couples magnetically with the VCO (see Figure 5.19). Two main magnetic coupling mechanisms are present: the on-chip coupling of inductors and the coupling between bond wires used to connect the die to the printed circuit board (PCB). The VCO is often realized as an LC-VCO. The inductor of the LC-VCO will couple magnetically with the inductor present in the last stage of a PA. This inductor is part of the output matching of the PA.

  Furthermore, the PA and the VCO are connected to a PCB through bond wires. Those bond wires also behave as inductors. As the bond wires are often close to be parallel wires, they can couple magnetically with each other.

- *Capacitive coupling*: The PA can couple capacitively with the VCO through their respective traces. Capacitive coupling can occur both between the on-chip interconnects and the interconnects on the PCB.

Different measurement experiments are set up to reveal which of the above-mentioned coupling mechanisms is the most important. The importance of the coupling mechanisms is obtained for a 4 GHz PPA and a 5–7 GHz LC-VCO [11].

5.4.1 Description of the Design of the PPA and the LC-VCO

This section describes the design of the prepower amplifier (PPA), which is the aggressor, and the LC-VCO, which is the victim. Both circuits show good performance. Therefore the circuits are representative candidates to study the different coupling mechanisms in a practical context.

5.4.1.1 The PPA as an Aggressor

The design of a fully integrated power amplifier in submicron CMOS technologies is challenging. The low breakdown voltage of the transistor limits the output power and the low substrate resistivity reduces the amplifiers power efficiency. Therefore only a PPA is realized in CMOS technology. This PPA drives the actual off-chip power amplifier, realized in alternative technologies like GaAs [12]. The latter technology offers a higher breakdown voltage and a higher substrate resistivity and therefore better performance.

A prepower amplifier operating at 4 GHz is designed in a UMC 0.13 μm CMOS triple-well technology on a lightly doped substrate (20 Ωcm). The PPA

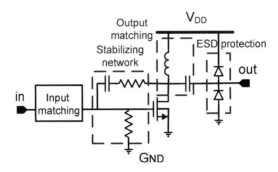

Figure 5.20   Schematic of the 4 GHz PPA.

consists of an NMOS power transistor loaded with an RF inductor of 0.8 nH (see Figure 5.20). This inductor serves to bias the drain of the transistor and is also part of an L-match circuit, which performs a conjugate match between the transistors' output impedance and the 50Ω load impedance. The power transistor is stabilized at high frequencies with a feedback RC network and at low frequencies with a 10 kΩ resistor to ground [13]. In this way, the power transistor is checked to be unconditionally stable up to 20 GHz. An input matching network is added for measurement purposes. The output of the PPA is ESD protected.

From the measurements shown in Figure 5.21(a), the output matching of the PPA is better then −10 dB from 3.25 GHz up to 4.1 GHz. Further, the PPA has a small-signal gain of 4 dB at 3.8 GHz. The input-referred 1 dB compression point is 0.5 dBm and the IIP3 is 13.7 dBm [see Figure 5.21(b)]. The PPA draws 17.8 mA from a 1.2V power supply when no input signal is applied.

5.4.1.2   Switched Varactor Bank LC-VCO as the Victim

The VCO under test is a 5–7 GHz LC-VCO. The gain block of the oscillator consists of a cross-coupled NMOS-PMOS transistor pair and an NMOS current mirror (see Figure 5.22). In this design, frequency tuning of the LC-VCO is obtained by changing the capacitance value of the resonant LC tank using multiple varactors [14, 15]. This means that a small varactor is used for the continuous fine tuning. The coarse discrete frequency changes are obtained by digitally switching the varactors. The use of multiple varactors results in a lower VCO gain, allowing an easier phase locked loop (PLL) design.

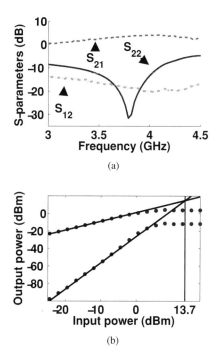

(a)

(b)

Figure 5.21 Performance measurements of the 4 GHz PPA. (a) The S-parameters of the 4 GHz PPA versus frequency. (b) The distortion. The IIP3 of the PPA is 13.7 dBm.

Compared to the VCO design described in Section 5.3.1, this VCO design is well shielded against substrate noise: The combination of a low VCO sensitivity [1], the triple well isolation, and a dedicated star shaped routing of the ground net makes the VCO very immune to coupling with the PPA. When the PPA carries no signal, the LC-VCO offers a worst case phase noise performance of −111 dBc/Hz at 1 MHz offset frequency [see Figure 5.23(b)] and a tuning range of 30% [see Figure 5.23(a)]. The VCO core draws 2.1 mA from a 1.2V power supply.

5.4.2 Coupling Mechanisms Between the PPA and the LC-VCO

When the PPA couples with the VCO, it causes sideband spurs in the spectrum of the VCO that are both modulated in amplitude and in frequency. Furthermore, the signal transmitted by the PPA will also couple to the output of the VCO, without

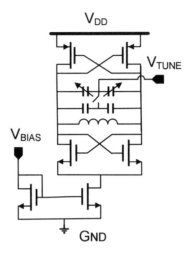

Figure 5.22   Schematic of the switched varactor LC-VCO.

frequency translation. Those different spurs can be seen in Figure 5.24. This figure shows the measured spectrum at the output of the VCO when it is operating at 5.1 GHz and when the PPA is excited with a sinusoidal signal of 3.68 GHz. The spurs are caused by different coupling mechanisms:

1. Resistively through the common substrate;

2. Magnetically between the on-chip inductors of the PPA and the LC-VCO;

3. Magnetically through the bonding wires of both circuits;

4. Capacitively between the traces of both circuits.

This section describes the different coupling mechanisms and gives an order of magnitude. Those estimations will be verified with measurements in the next section.

5.4.2.1   Resistive Coupling Through the Common Substrate

When the PPA amplifies the RF signal, it draws high current peaks from its supply lines. This results in ringing on the supply lines. This ringing is injected into the substrate by a resistive coupling of the ground node of the PPA via substrate contacts. It then propagates through the common substrate as substrate noise.

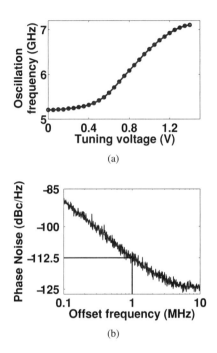

(a)

(b)

Figure 5.23 Performance measurements of the 5–7 GHz LC-VCO. (a) The tuning range of the LC-VCO is 30 %. (b) The worst case phase noise performance is -111 dBc/Hz at 1 MHz offset.

To investigate the behavior of this coupling via the substrate, a dedicated substrate contact of 114 μm by 58 μm has been placed next to the VCO. To measure the transfer function of the coupling, a large sinusoidal signal, whose frequency is swept between 10 MHz and 250 MHz, is injected through this contact. The established power on the RF source is 14 dBm to guarantee enough measurement accuracy. The reflection coefficient of the substrate is −6 dB. To minimize reflections in the cables of the bias lines, bias tees are used. The power of the sideband spurs is measured with a spectrum analyzer (HP8565E). Figure 5.25 shows the measured right sideband spur. The coupling mechanisms are similar to those obtained in Section 5.3.2. For low frequencies, substrate noise coupling results in narrowband FM modulation of the LO signal. The spurs are decreasing at a rate of 20 dB per decade with increasing offset frequency. Starting from 70 MHz, the spurs increase at a rate

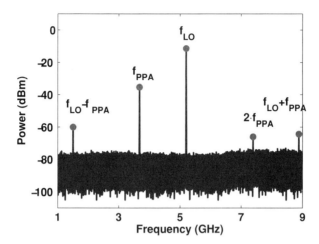

Figure 5.24   Spectrum of the VCO operating at 5.1 GHz when the PPA is excited at 3.68 GHz.

of 20 dB per decade with increasing offset frequency. Those spurs are modulated in amplitude.

5.4.2.2   Magnetic Coupling Between the On-Chip Inductor of the PPA and the LC-VCO

The on-chip inductor of the LC-VCO and the RF choke of the PPA act as a transformer. The mutual coupling will cause a voltage change in the LC tank of the VCO. This superimposed voltage results in modulated sideband spurs and spurs without frequency translation. The mutual inductance, simulated using the ASITIC software [16], is in this design 3.1 pH. Figure 5.26 shows the simulated mutual inductance versus the distance (d) that is measured between the centers of the inductors.

5.4.2.3   Magnetic Coupling Between the Bonding Wires of the PPA and the LC-VCO

The chip containing the PPA and the VCO is mounted on a PCB. Consequently they are connected with bond wires to the PCB traces. The bond wires of the PPA will

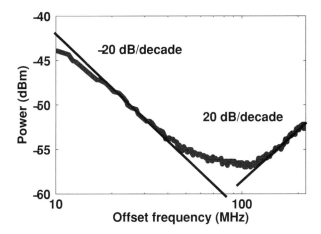

Figure 5.25 Right sideband spur measured with a spectrum analyzer when the VCO is oscillating at 5.4 GHz.

also couple magnetically with the bond wires of the VCO. A sensitivity analysis is performed to determine which connection of the LC-VCO to the PCB is the most sensitive to bond wire coupling. This connection will cause the largest degradation on the performance of the LC-VCO. SpectreRF [17] simulations show that the VCO is the most sensitive to bond wire coupling into the V_{DD} connection of the VCO. This can be explained by the absence of a tail transistor on top of the cross-coupled NMOS/PMOS pair (see Figure 5.22).

The mutual inductance between the closest PPA bond wire (output of the PPA) and VCO bond wire (V_{DD} of the VCO) is computed using FastHenry software [18]. The bond wire is 25 μm thick and approximately 1 mm long. The mutual inductance between both bond wires is as large as 205.7 pH. Figure 5.27 shows the simulated mutual inductance versus the lateral distance (d) between the bond wires.

The simulations of Figures 5.26 and 5.27 show that the mutual inductance between bond wires is almost two orders of magnitude larger then the mutual inductance between the on-chip inductors. However, one has to keep in mind that the amplitude of the spurs does not only depend on the mutual inductance but also on the sensitivity of the node.

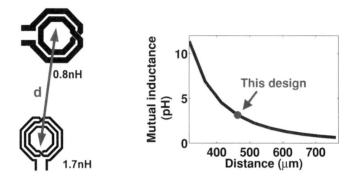

Figure 5.26   ASITIC simulations of the mutual inductances between the RF choke of the PPA and the inductor of the VCO versus the center distance between both inductors.

5.4.2.4   Capacitive Coupling

Capacitive coupling can occur between the on-chip traces as well as between the PCB traces. The PPA and the LC-VCO are routed separately from each other both on-chip and on PCB. Hence, the worst case capacitive coupling will occur between the closest PCB trace of the PPA (output of the PPA) and the VCO (V_{DD} of the VCO). Those two traces are simulated with the HFSS field solver [19]. The resulting S-parameters are used together with the FastHenry [18] model of the bond wire to model the behavior of the full circuit with SpectreRF [17]. Figure 5.28 shows that the largest capacitive coupling between the PCB traces is still 30 dB lower at 4 GHz than the magnetic coupling between the bond wires.

A parasitic extraction has been performed on the layout of the chip. This reveals that the worst-case coupling capacitance between the on-chip traces is smaller than 0.02 fF. Therefore, capacitive coupling can be neglected.

5.4.3   Measuring the Dominant Coupling Mechanisms

In this section different experiments are carried out to reveal the dominant coupling mechanisms. First, the measurement setup is discussed. Next, dedicated experiments are used to measure the impact of the resistive substrate coupling, the magnetic coupling between the on-chip inductors, and the bond wires.

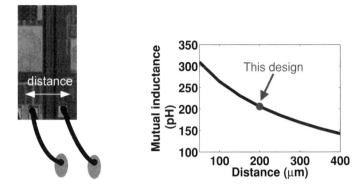

Figure 5.27 FastHenry simulations of the mutual inductance of the bond wire of the output of the PPA and the V_{dd} of the VCO versus the distance between both bond wires.

5.4.3.1 Measurement Setup

The die, consisting of the PPA and the VCO (see Figure 5.29), is wire-bonded on a single PCB. The PPA and the VCO have separate on-chip supply lines and those lines are also kept separated on PCB to avoid capacitive coupling. Both circuits are fed with a different voltage source. The transfer function from the input of the PPA to a single output of the VCO is measured with a network analyzer (HP8753ES). The modulated sideband spurs are measured with a spectrum analyzer (HP8565E). The transfer function and the sideband spurs are both measured when the PPA is excited from 3.5 GHz up to 4.1 GHz because the PPA is matched at the output in that frequency region. One single output of the VCO and the output of the PPA are terminated with a 50Ω impedance. The injected power is −2 dBm to have enough measurement accuracy. The output power of the PPA is 1 dBm at 3.8 GHz.

5.4.3.2 Impact of Substrate Coupling

The impact of the coupling through the substrate can be determined by physically separating the PPA from the VCO. This is done by dicing both circuits (see Figure 5.30). For that purpose die seals are foreseen in the design. Those die seals prevent damage to the circuits due to cracks that appear during the dicing process [20]. The performance of both circuits is checked after dicing. No performance degradation is observed. After dicing, both circuits are separated by a 30 μm air gap. This air gap acts as a capacitor whose value is sufficiently small. This assumption is

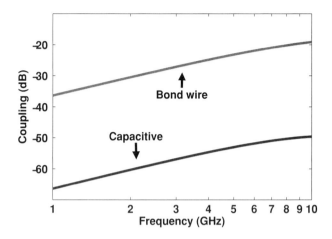

Figure 5.28   Magnetic coupling between the bond wires dominates over the capacitive coupling between the PCB traces.

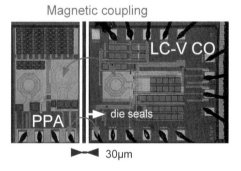

Figure 5.29   After dicing, coupling between the bond wires of output of the PPA and the V_{DD} of the VCO determines the power of the unwanted spurs.

validated in the next section. As the width of the air gap is very small, the physical distance between the inductors and the bond wires remains the same. Thus, the magnitude of the magnetic coupling is almost not affected.

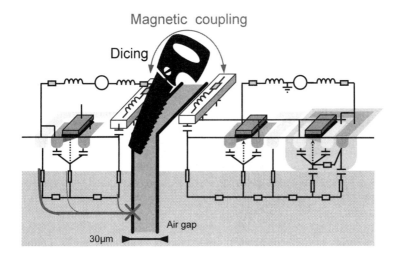

Figure 5.30 Dicing the PPA from the VCO removes the substrate coupling.

Figure 5.31 compares the transfer function from the input of the PPA to a single output of the VCO before (undiced) and after the dicing (diced 30 μm). The spurs are attenuated by 8 dB at 3.75 GHz. This significant attenuation shows that the coupling through the substrate is the dominant coupling mechanism. However it also shows that magnetic coupling cannot totally be neglected as it is only a factor of three lower than the resistive coupling.

The dicing operation clearly demonstrates that the coupling through the substrate is the dominant coupling mechanism. However, practical applications do not allow this dicing operation. Therefore, P$^+$ diffused guard rings are often used to reduce the substrate coupling.

5.4.3.3 Impact of Magnetic Coupling Between the On-Chip Inductors

According to EM simulations, the magnetic coupling between the on-chip inductors can be strongly reduced by increasing the distance between the PPA and the VCO (see Figure 5.26). Therefore, the PPA was moved 200 μm away from the VCO (see Figure 5.32). According to simulations done with ASITIC, the mutual inductance

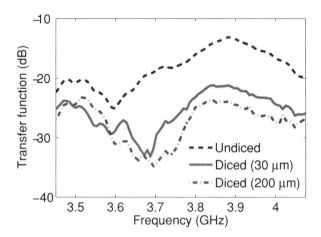

Figure 5.31   Measured transfer function from the input of the PPA to the output of the VCO.

is then reduced to 0.99 pH instead of 3.1 pH. Thus the expected mutual coupling between the on-chip inductors is lowered by 10 dB.

The difference in coupling between the bond wire is expected to be only 1 dB lower. Due to the displacement, the bond wires of the PPA are rotated over a small angle. Simulations done with FastHenry predict that this rotation decreases the mutual inductance between the bond wire of the output of the PPA and the V_{DD} of the VCO with only 1 dB.

Moreover, the residual substrate coupling over the air gap will also be affected by this displacement. Since the capacitance over the air gap is inversely proportional with the width of the air gap, the displacement from 30 μm to 200 μm results in a further attenuation of 16 dB.

Figure 5.31 shows that increasing the distance between the PPA and the VCO only reduces the spurs by 2 dB. Therefore it can be concluded that bond wire coupling is more important than magnetic coupling between the on-chip inductors, and that the capacitance over the air gap can indeed be neglected.

5.4.4   Conclusions

In this section, we have analyzed the coupling of a transmitted PPA signal into a state-of-the-art shielded LC-VCO. The transmitted signal of the PPA causes

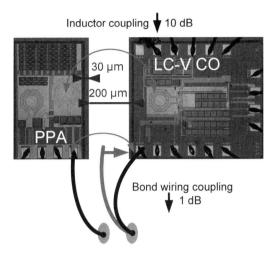

Figure 5.32 The PPA is moved 200 μm away from the VCO in order to reveal which type of magnetic coupling is more important.

unwanted spurs to appear in the spectrum of the VCO. The PPA can couple in different ways with the VCO, namely:

- Resistively through the substrate;
- Capacitively through the traces of both circuits;
- Magnetically through the bonding wires of both circuits;
- Magnetically through the on-chip inductors of both circuits.

The different coupling mechanisms are discussed and quantitative information is given. Measurements of the transfer function between a dedicated substrate contact and the output of the VCO point out that the substrate noise behavior is similar to the results obtained in Section 5.3.2. Furthermore, the mutual inductance between the bonding wires and the on-chip inductors of the PPA and the VCO is estimated. Those estimations show that the mutual inductance between the bonding wires is almost two orders of magnitude higher than the mutual inductance between the on-chip inductors. Further simulations show that the largest capacitive coupling between the PCB traces is still 30 dB lower than magnetic coupling between the bonding wires. Therefore capacitive coupling can be neglected.

Dedicated experiments reveal the dominant coupling mechanism between the PPA and the VCO. Dicing both circuits shows that the coupling through the common substrate is the dominant coupling mechanism. Enlarging the air gap between both circuits points out that the remaining coupling is mainly due to the magnetic coupling between the bond wires.

5.5 CONCLUSIONS

Measurements are shown to provide reliable information about the noise coupling in general, but they cannot reveal how noise couples into the circuit because the internal nodes of the circuit cannot be accessed. Simulations are mandatory to uncover how noise enters the analog/RF circuit and how it propagates through this circuit. For aggressor cases, which are related to digital switching noise and PA coupling, the substrate plays a primary role. Therefore, in order to *predict* the dominant noise coupling in analog/RF circuits, it is necessary to build a methodology that includes a model for the substrate. The modeling of the substrate noise impact on analog/RF circuit is the subject of Chapter 6. Note that the measurements in this chapter have clearly demonstrated that careful modeling of the assembly characteristics and the layout details of both the chip and the PCB is required to bring simulations and measurements in agreement. Once the simulation model is able to explain the measured behavior, simulations can provide more information about the noise coupling mechanisms. Only then can countermeasures be proposed to increase the immunity of the circuit against substrate noise.

References

[1] C. Soens, G. Van der Plas, P. Wambacq, S. Donnay, and M. Kuijk, "Performance degradation of LC-tank VCOs by impact of digital switching noise in lightly doped substrates," *IEEE Journal of Solid States*, Vol. 40, No. 7, July 2005, pp. 1472–1481.

[2] M. Mendez, D. Mateo, X. Aragones, and J. Gonzalez, "Phase noise degradation of LC-tank VCOs due to substrate noise and package coupling," *Proc. 31st European Solid-State Circuits Conference ESSCIRC 2005*, September 12–16, 2005, pp. 105–108.

[3] N. Checka, D. Wentzloff, A. Chandrakasan, and R. Reif, "The effect of substrate noise on VCO performance," *Proc. Digest of Papers Radio Frequency Integrated Circuits (RFIC) Symposium 2005 IEEE*, 2005, pp. 523–526.

[4] G. Brenna, D. Tschopp, J. Rogin, I. Kouchev, and Q. Huang, "A 2-GHz carrier leakage calibrated direct-conversion WCDMA transmitter in 0.13 μm CMOS," *JSSC*, Vol. 39, No. 8, August 2004, pp. 1253–1262.

[5]   S. Bronckers, G. Vandersteen, C. Soens, G. Van der Plas, and Y. Rolain, "Measurement and modeling of the sensitivity of LC-VCO's to substrate noise perturbations," *Proc. IEEE Instrumentation and Measurement Technology*, 2007, pp. 1–6.

[6]   C. Soens, G. Van der Plas, P. Wambacq, and S. Donnay, "Simulation methodology for analysis of substrate noise impact on analog/RF circuits including interconnect resistance," *Proc. Design, Automation and Test in Europe*, Vol. 1, 2005, pp. 270–276.

[7]   MATLAB documentation, http://www.mathworks.com.

[8]   M. Badaroglu, G. Van der Plas, P. Wambacq, S. Donnay, G. Gielen, and H. De Man, "SWAN: high-level simulation methodology for digital substrate noise generation," *Symposium on VLSI Circuits*, Vol. 14, No. 1, January 2006, pp. 23–33.

[9]   M. Badaroglu, M. van Heijningen, V. Gravot, S. Donnay, H. De Man, G. Gielen, M. Engels, and I. Bolsens, "High-level simulation of substrate noise generation from large digital circuits with multiple supplies," *Proc. Design, Automation and Test in Europe Conference and Exhibition 2001*, March 13–16, 2001, pp. 326–330.

[10] S. Donnay and G. Gielen, *Substrate Noise Coupling in Mixed-Signal IC's*, Boston, MA: Kluwer Academic Publishers, 2003.

[11] S. Bronckers, G. Vandersteen, L. De Locht, G. Van der Plas, and Y. Rolain, "Study of the different coupling mechanisms between a 4 GHz PPA and a 5-7 GHz LC-VCO," *Proc. IEEE Radio Frequency Integrated Circuits Symposium RFIC 2008*, 2008, pp. 475–478.

[12] S. Cripps, "RF power amplifiers for wireless communications," *Microwave*, Vol. 1, No. 1, March 2000.

[13] G. Gonzalez, *Microwave Transistor Amplifier Analysis and Design*, Upper Saddle River, NJ: Prentice-Hall, 1984.

[14] K. Manetakis, D. Jessie, and C. Narathong, "A CMOS VCO with 48% tuning range for modern broadband systems," *Proc. Custom Integrated Circuits Conference the IEEE 2004*, October 3–6, 2004, pp. 265–268.

[15] D. Hauspie, E.-C. Park, and J. Craninckx, "Wideband VCO with simultaneous switching of frequency band, active core, and varactor size," *JSSC*, Vol. 42, No. 7, July 2007, pp. 1472–1480.

[16] A. Niknejad, "Modeling of passive elements with ASITIC," *Proc. IEEE Radio Frequency Integrated Circuits (RFIC) Symposium*, June 2–4, 2002, pp. 303–306.

[17] Spectre RF, http://www.cadence.com/products/custom_ic/spectrerf.

[18] C. Schuster, G. Leonhardt, and W. Fichtner, "Electromagnetic simulation of bonding wires and comparison with wide band measurements," *ADVP*, Vol. 23, No. 1, February 2000, pp. 69–79.

[19] HFSS, http://www.ansoft.com/products/hf/hfss/.

[20] A. V. Kearney, A. V. Vairagar, H. Geisler, E. Zschech, and R. H. Dauskardt, "Assessing the effect of die sealing in Cu/Low-k structures," *Proc. International Interconnect Technology Conference IEEE 2007*, June 4–6, 2007, pp. 138–140.

Chapter 6

The Prediction of the Impact of Substrate Noise on Analog/RF Circuits

6.1 INTRODUCTION

The previous chapter exploited different measurement techniques to obtain a large amount of information about the different substrate noise coupling mechanisms. Those measurement techniques can only be used once the circuit is realized. Moreover, measurements cannot reveal how substrate noise couples into the circuit and how it propagates toward the output of the circuit, since the internal nodes cannot be accessed during the measurements. Therefore this chapter focuses on simulation methodologies to predict the impact of substrate noise before the chip is realized. Two simulation methodologies are proposed. The first methodology uses the finite difference method (FDM) to model the substrate. The FDM method is implemented by the tool SubstrateStorm [1]. The second methodology uses the finite element method (FEM) to characterize the substrate. The FEM method is implemented by HFSS [2].

The two methodologies are demonstrated on an analog/RF circuit. The circuits under test are chosen differently in order to show that the substrate noise coupling mechanisms are not necessarily the same, although the circuit topology is similar. Chapter 4 already pointed out that at low frequencies, where capacitive and inductive effects can be neglected, substrate noise couples either through the bulk or the ground of the transistors. Through simulations and corresponding measurements it will be shown that this statement is also valid for complete analog/RF circuits.

Both simulation methodologies provide better insight into the different noise coupling mechanisms and help the designer take the appropriate countermeasures to increase the immunity of his or her analog circuits.

Layout and circuit techniques are proposed, which can significantly reduce the noise coupling and increase the success rate of a first time right silicon pass. Finally, it is shown that 3D stacking offers a huge opportunity to reduce the substrate noise coupling mechanisms.

6.2 THE SUBSTRATE MODELED WITH FDM

The substrate is modeled here as a network consisting of lumped elements only. First, a simulation model is built that incorporates the dominant substrate coupling mechanisms. Second, we will illustrate how this approach can be used to investigate the impact of substrate noise in analog/RF circuits. To make this more practical, the approach is applied to a 900 MHz LC-VCO that is perturbed by a substrate noise signal. The substrate coupling mechanisms are revealed with simulations from DC up to LO frequency. Afterward the simulations are validated with measurements.

6.2.1 Impact Simulation Methodology

In order to reveal the substrate coupling mechanisms, the model needs to incorporate the dominant coupling mechanisms. Since the substrate noise coupling mechanisms are influenced by the on-chip layout details and the parasitics of the PCB, a simulation model is built up that consists of an on-chip model and an off-chip model. The on-chip model includes the RF models of the devices, the parasitics of the interconnects, and the substrate. The off-chip model includes the parasitics of the PCB components and the interconnects. This section describes the on- and off-chip simulation model. The complete workflow of the simulation methodology is given in Figure 6.1. The next section applies this impact simulation methodology to an example, which is a 900 MHz LC-VCO.

6.2.1.1 On-Chip Simulation Model

The extraction of the on-chip simulation model starts from the layout of the analog/RF circuit. The circuit netlist containing models for the devices and the parasitics of the interconnects are extracted. To this end an extraction deck is written

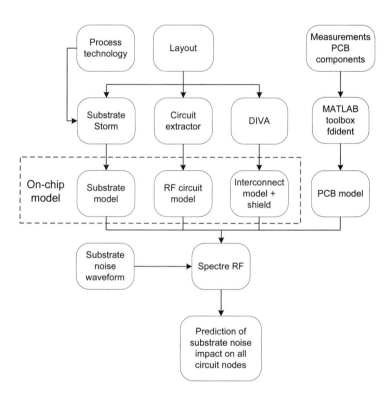

Figure 6.1   Substrate noise simulation methodology.

in DIVA [3]. The parasitic resistances of the interconnects are determined based on the geometry and the sheet resistance of the interconnect. The capacitance of the interconnect is determined based on its geometry and the dielectric properties of the oxide. The interconnects are modeled as an RC mesh in the circuit netlist. It is important to note that those capacitors are referred to the ideal ground. Hence, substrate noise cannot couple capacitively from the substrate into the interconnects. Including the capacitive coupling from the substrate to each metal interconnect is too CPU intensive and not feasible. Therefore only the capacitive coupling to large metal structures such as inductors is taken into account.

The DIVA deck recognizes all the devices in the layout and their RF models are added to the circuit netlist.

Next, SNA [4] is used to extract an RC netlist for the substrate based on doping profile information. Those doping profiles need to be provided by the foundry and are unique for each technology. The resulting substrate netlist contains two types of external nodes. The first type is the nodes for the connections of the die to the devices in the circuit. The second type is the substrate contacts used to inject disturbances in the substrate or the noisy nodes of a digital circuit.

Finally, the circuit netlist and substrate netlist are connected and merged into one single on-chip simulation model.

6.2.1.2   PCB Simulation Model

In order to build a simulation model that can predict the substrate noise behavior over a broad frequency range, a model for the PCB needs to be built. Chapter 5 revealed that the substrate noise behavior is influenced by the different devices on the PCB and the bonding wire, which connects the chip to the PCB. Therefore, a model is built for the devices on the PCB, the bonding wires, and the PCB traces.

The PCB devices can be easily modeled with measurements because they are readily available in an IC design house. The different devices on the PCB are measured with an impedance analyzer. The impedance analyzer returns the impedance value Z as a function of frequency. This impedance function $Z(s)$ is then modeled as a linear transfer function in the Laplace variable s [5]. This results in a rational function in s, which is identified by the MATLAB toolbox called fdident [6]. The coefficients of this transfer function are used to determine the RLC values of the corresponding lumped network.

As an example, the modeling of PCB decoupling capacitors will be demonstrated. Decoupling capacitors are always present on the PCB to decouple the bias

lines. Those bias lines are generally decoupled off-chip with surface mounted devices (SMD) capacitors with a set of different values (e.g., 100 pF, 100 nF, and
100 μF) in order to decouple in a broad frequency range [7]. After measuring the
capacitor, the measurements are modeled as rational function in s by a frequency
domain estimator. In the case of a 100 nF SMD decoupling capacitor, this gives:

$$Z(s) = \frac{2.6 \cdot 10^{-17} \cdot s^2 + 1.7 \cdot 10^{-9} \cdot s + 1}{8.8 \cdot 10^{-8} \cdot s} \tag{6.1}$$

This transfer function corresponds to a series connected RLC network:

$$Z(s) = \frac{L \cdot C \cdot s^2 + C \cdot R \cdot s + 1}{C \cdot s} \tag{6.2}$$

From (6.1) and (6.2), the RLC values can be determined (see Figure 6.2).

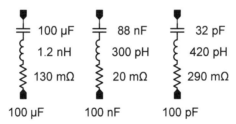

Figure 6.2   The decoupling capacitors are modeled by a frequency domain estimator and mapped into
an equivalent RLC network.

Simulations of this equivalent network reveal that the mean error between the
model and the measurements is less than 1 dB in the frequency range from 1 MHz
to 1 GHz (see Figure 6.3). The absolute error increases for higher frequencies due
to higher order poles which are not taken into account in our lumped model.
    The PCB traces and the bond wires are modeled as transmission lines. The
values of the inductance and resistance of the bonding wires and the PCB traces
are determined based on their geometry [8]. The models of the bonding wires, PCB
traces, and PCB decoupling capacitors are merged into one single PCB simulation
model.

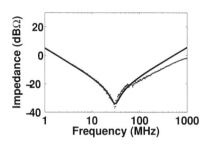 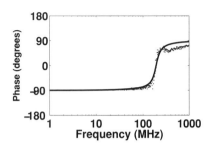

Figure 6.3   The 100 nF SMD decoupling capacitors is measured and modeled as a linear transfer function.

6.2.2   Prediction of the Impact of Substrate Noise from DC Up to LO Frequency

The simulation methodology is modeled in a high-ohmic 0.18 μm 1P6M CMOS technology and is applied on the 900 MHz LC-VCO design that is used in Chapter 5 to demonstrate some measurement techniques that enabled one to extract a large amount of information about the substrate noise coupling mechanisms. The schematic of this VCO is shown in Figure 6.5. The oscillation frequency of the 900 MHz LC-VCO is determined by the resonance frequency of the LC tank. The losses of the LC tank are compensated by the NMOS/PMOS cross-couples pair. The current through the VCO core is determined by the PMOS current source. As already described in Chapter 5, this VCO offers an acceptable performance.

The extraction of the model of the substrate and the interconnects takes less than 1 hour on a HPUNIX server. This short simulation time could only be obtained by neglecting most of the capacitive coupling from the substrate to the interconnects. Only the capacitive coupling from the shield of the inductor that is embedded in the substrate to the inductor itself is taken into account. Including the capacitive coupling to every interconnect would increase the simulation time drastically up to more than 100 hours.

The waveform resulting from substrate noise impact can be predicted on all the nodes of the circuit. The simulated sideband spurs at the output of the VCO are shown in Figure 6.4. The behavior of those spurs is similar to the measured spurs in Section 5.3.2. Section 5.3.2 divided the substrate noise behavior into four frequency regions:

1. Sideband spurs caused by low frequency perturbations result in FM spurs around the LO. The spurs are decreasing at a rate of 20 dB/decade.

2. At intermediate frequencies the dominant impact mechanism is moving from FM toward AM spurs. The behavior of the spurs changes from a decrease at a rate of 20 dB/decade to an increase at a rate of 20 dB/decade and later on even a rate of 40 dB/decade.

3. AM spurs are dominant at high frequencies. The spurs increase at a rate of 40 dB/decade.

4. Close to the LO frequency, pulling and locking of the LO determine the behavior of the perturbation.

   Simulations will now provide more insight into the different coupling mechanisms because the internal nodes of the circuit can now be accessed. This section reveals how substrate noise couples into the VCO and how the substrate noise signal is converted into FM and AM modulated spurs. The next section will compare the measurements performed in Section 5.3.2 with the corresponding simulations performed in this section.

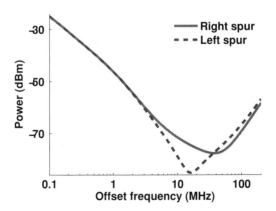

Figure 6.4   Simulated left and right spur of the 900 MHz LC-VCO.

6.2.2.1 Impact at Low Frequencies

From DC to 10 MHz the substrate noise signal couples resistively into the nonideal ground and causes ground bounce. This is in agreement with Chapter 4, which stated that when the resistance of the ground interconnect is larger than 0.65Ω, ground bounce dominates. For values of the ground resistance lower than 0.65Ω, the bulk effect dominates. In this case, the resistance of this nonideal ground is around 7Ω and hence ground bounce dominates. Ground bounce causes a voltage fluctuation at the source of the NMOS transistors (see Figure 6.5) and modulates the voltage across the variable capacitances. The variable capacitances present in this design are the varactors of the LC-VCO and the parasitic capacitances of the NMOS and PMOS transistors. The modulation of those capacitances results in narrow band FM of the LO signal. The spurs are decreasing at a rate of 20 dB/decade with the offset frequency [9] (see Figure 6.4).

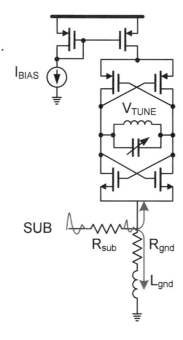

Figure 6.5   At low frequencies substrate noise couples into the nonideal ground of the LC-VCO.

6.2.2.2   Impact at Intermediate Frequencies

At intermediate frequencies (10 MHz to 100 MHz) the amplitude of the spurs no longer decreases by 20 dB/decade but starts to increase by 20 dB/decade. A part of the substrate noise signal is drained resistively by the nonideal on-chip ground toward the PCB through the ground bond wire. Another part of the substrate noise signal couples resistively in the shield of the inductor. This shield is connected to the PCB with a dedicated bond wire. Part of the signal picked up by the shield is drained toward that dedicated bond wire. Another part of the signal flows through the parasitic capacitance of the inductor (Figure 6.6). Hence the substrate noise signal couples capacitively to the inductor. This signal is superposed on the oscillation signal across the LC tank. Due to the PMOS current source, the voltage swing of the VCO is limited by the current that is delivered by this current source. Thus, the superposition of the substrate noise currents that couple in to the inductor of the VCO change the voltage swing of the VCO and modulate the LO in amplitude. Capacitive coupling resulting in AM-modulated spurs explains the increase of 20 dB/decade.

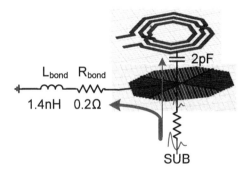

Figure 6.6   Starting from 10 MHz, the substrate noise signal couples capacitively through the inductor.

6.2.2.3   Impact at Higher Frequencies

At higher frequencies the substrate noise impact increases by 40 dB/decade. Above 100 MHz the impedance of the shield toward the PCB is determined by the impedance of the bonding wire connected to the shield. This can easily be shown by removing the capacitive coupling to the inductor and resimulating the simulation model. Figure 6.7 shows that without capacitive coupling the spurs increase at a

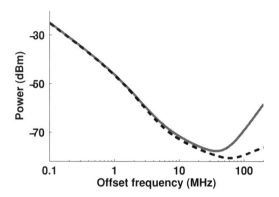

Figure 6.7 The solid line reflects the simulation of the left sideband spur with capacitive inductor coupling. The dotted line reflects the simulation of the left sideband spur without capacitive inductor coupling.

rate of 20 dB/decade instead of 40 dB/decade. A larger part of the signal couples capacitively to the inductor and less signal is drained to the PCB. The capacitive coupling to the inductor together with the inductance of the bonding wire acts as a second order system. This explains the increase of the substrate noise impact at a rate of 40 dB/decade. The dominant impact mechanism above 100 MHz is AM modulation. DC decouple capacitors at the output of the VCO will attenuate the left sideband spurs resulting in an injected substrate noise signal above 650 MHz (see Figure 6.8) .

6.2.2.4   Impact Close to the LO Frequency

At an offset frequency of 150 MHz from the LO frequency, substrate noise starts pulling and locking effects dominate the substrate noise behavior. Pulling and locking are nonlinear dynamical phenomena and are not included in the presented model. However nonlinear macromodels can be found in [10].

6.2.3   Experimental Validation of the Simulation Methodology

This section compares the simulations of the previous section with the corresponding measurements. The previous section already pointed out that the simulated substrate noise behavior is similar to the measured one. The goal of this comparison is to show how accurate the proposed methodology can predict the power of the spurs caused by substrate noise coupling.

6.2.3.1   Validation of the Different Coupling Mechanisms

The VCO is mounted on a PCB. A large sinusoidal signal is injected into the substrate through a dedicated substrate contact. The size of the substrate contact is 10 μm by 20 μm. The established power on the RF source is 10 dBm to guarantee enough measurement accuracy as before.

The power of all the spurs (the direct coupled spur and the spurs resulting from mixing of the injected signal with the first and second harmonic of the LO) are measured with the spectrum analyzer. The frequency of the injected signal is varied from 1 MHz up to LO frequency while the tuning voltage is varied from 0V up to 1.8V.

In order to validate the coupling mechanism of substrate noise described in Section 6.2.2, the bias lines are not decoupled on PCB.

Figure 6.8 shows a good agreement between the measured spurs and simulation. The mean error between measurements and simulations is less than 3 dB.

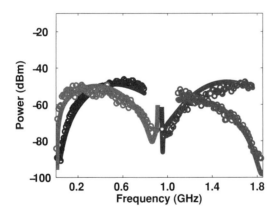

Figure 6.8   The power of the direct coupled spur, left sideband spur of the first and second harmonic are measured with a spectrum analyzer for a tuning voltage of 0.9V. The *o* represents the measurement and the solid lines reflect the simulations.

In Section 5.3.4 a dedicated measurement was set up to show that above 100 MHz the substrate noise impact is determined by the impedance of the bonding wire that is connected to the shield of the inductor. Different PCBs with different contact resistances of the bonding wires were measured in order to show the influence of this resistance on the impact of substrate noise. Measurements pointed out that the PCB with a DC resistance of the bonding wire that is 6.6 times larger resulted in 6 dB higher impact. Figure 6.9 shows that this effect is also predicted by simulations.

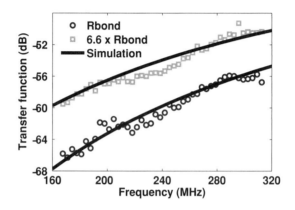

Figure 6.9   The impedance of the bonding wire and the on-chip ground resistance plays a major role in the substrate noise impact mechanism.

6.2.3.2   Validation of the Influence of the PCB Decoupling Capacitors

As depicted in Section 5.3.5, the PCB decoupling capacitors influence the substrate noise coupling at high frequencies (above 100 MHz). Decoupling the bias lines with a 100 μF decoupling causes a resonance at 480 MHz. This resonance is because the PCB decoupling capacitor behaves as an inductor at those frequencies and resonates with the on-chip decoupling capacitors. Figure 6.10 shows that the accurate modeling of the PCB and the decoupling capacitor resulted in an accurate estimate of the transmission zero of the transfer function.

6.2.4   Conclusions

This section uses the FDM method to characterize the substrate for the prediction and analysis of substrate noise on analog/RF circuits. The FDM method is implemented by the tool SubstrateStorm. The tool enables one to reveal the dominant

substrate noise coupling mechanisms in an 900 MHz LC-VCO. At low noise frequencies (< 10 MHz) substrate noise causes ground bounce. Since the resistance of
the ground interconnect is larger than 0.65Ω, the bulk effect can safely be neglected.
Ground bounce results in FM modulated spurs. At high frequencies (> 100 MHz),
substrate noise couples capacitively into the on-chip inductor and causes AM modulated spurs. The methodology is successfully validated with measurements.

Hence, SubstrateStorm, which implements the FDM method, is a powerful
tool to analyze the impact of substrate noise on analog/RF circuits; however, it
has a serious number of limitations. The necessity of known doping profiles and
the limited capability to model the capacitive coupling from the substrate to metal
interconnects is a serious disadvantage of the tool. The abovementioned limitations
are circumvented in the next section, which uses the FEM method to predict the
impact of substrate noise in analog/RF circuits. The FEM method is implemented
by the tool HFSS [2]. This tool does not need doping profiles, is easier to use, and
is as accurate as the SubstrateStorm simulation.

6.3 SUBSTRATE MODELED BY THE FEM METHOD

This methodology circumvents the shortcomings of the methodology of the previous section by combining the strengths of the EM simulator (like HFSS [2] or
Momentum [11]) and the circuit simulator (like SpectreRF [12]). Here, the substrate
and the interconnects are characterized by an EM simulator and are thus represented by a finite element model. The finite element model is represented by an S-
parameter matrix. This S-parameter matrix is simulated together with the RF models
of the active devices. The resultant waveforms provided by the circuit simulator,
gives the designer insight in the dominant coupling mechanisms. To make this more
practical, the substrate noise behavior is predicted and the substrate coupling mechanisms are revealed in an example. Here, the example is a millimeter-wave VCO
that operates at frequencies between 48 GHz and 53 GHz. The predicted substrate
noise behavior is afterward verified with measurements on a real-life prototype of
the millimeter-wave VCO.

6.3.1 Impact Simulation Methodology

The simulation methodology is the same as used in Chapter 4 to analyze the
different coupling mechanisms in an active device. The simulation methodology
consists of two simulations. First, an EM simulation is performed. This simulation

includes the on-chip interconnects, the substrate, and the passive components. Since the substrate noise signals are sufficiently small, the interconnects and the substrate can be considered to behave linearly. Next, a simulation model is built to fully characterize the analog/RF circuit: the results of the EM simulation are used together with the RF models of the devices to perform a circuit simulation. The resulting waveforms that are present on the different terminals of the simulation model will give the designer insight into the different substrate noise coupling mechanisms. A complete block diagram of the methodology is given in Figure 6.11. This section briefly discusses which user interaction is required to properly set up the different simulations in order to predict the impact of substrate noise. The next section applies this simulation methodology to an example.

6.3.1.1 EM Simulation

This step starts from the layout of the analog/RF circuit. First, the layout is simplified. This simplified layout is then streamed into the EM simulator. Next, the EM environment is set up and ports are placed. Finally, this EM environment is properly set up and simulated. The transformation of the layout consists of a number of steps that are described below.

- *Simplifying the layout:* It is mandatory to simplify the layout to speed up the EM simulations. To eliminate as many details as possible without jeopardizing the accuracy of the result, the following rules are used:

  - The transistors are removed from the layout since they can be represented by their respective RF model.

  - The different vias that connect the different metal layers are grouped.

  - The ground shield, which is perforated to meet the stringent DRC rules, is filled with metal.

     This simplified layout is then imported into the EM environment. This environment includes a substrate of 20 Ωcm, a P-well of 800 S/m, the silicon dioxide with an ϵ_r of 3.7, and an air box on top of the chip.

- *Placing ports:* All the transistors are removed and are replaced by three ports as explained in Chapter 4. Further, ports are placed at the external connections of the analog/RF circuit and at the substrate contact. This substrate contact is used to inject substrate noise into the substrate and hence replaces the switching digital circuitry in this experiment.

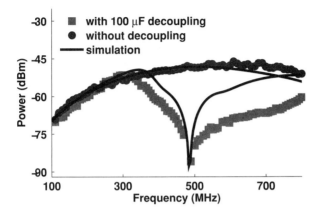

Figure 6.10 Simulation versus measurement of the direct coupled spur with and without decoupling the PCB traces with a capacitor of 100 μF.

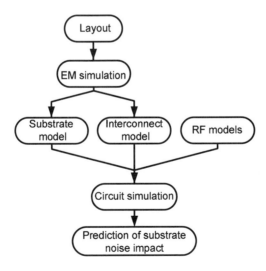

Figure 6.11 Impact simulation approach.

• The S-parameters are solved at the different ports of the EM environment for a user-given frequency range and accuracy.

6.3.2 Prediction of the Impact of Substrate Noise

In order to predict and understand the different coupling mechanisms, it is mandatory that the designer knows and understands how the analog/RF circuit is designed. This section briefly describes the analog/RF circuit under test. The analog/RF circuit under test is a 48–53 GHz LC-VCO that is designed in a UMC 0.13 μm CMOS technology. The gain block of the oscillator consists of a cross-coupled NMOS transistor pair and an NMOS current mirror (see Figure 6.12). NMOS transistors are preferred over PMOS transistors because they exhibit a lower parasitic capacitance for the same transconductance value. In this way, a higher tuning range can be achieved. The frequency tuning of the LC-VCO is obtained by changing the capacitance value of the resonant LC-tank. The tunable capacitors are implemented as NMOS transistors whose drain and source nodes are shorted. The inductor of the LC-tank is implemented as a folded microstrip line [13]. The VCO is buffered by source followers.

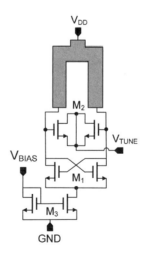

Figure 6.12   Schematic of the millimeter-wave LC-VCO.

The LC-VCO offers a phase noise performance of −84.2 dBc/Hz at 1 MHz offset frequency [see Figure 6.13(a)] and a tuning range of 10% [see Figure 6.13(b)]. The VCO core draws 4.5 mA from a 1V power supply.

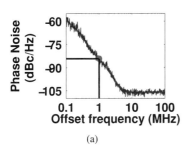
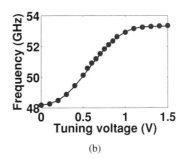

(a)   (b)

Figure 6.13   Measured performance of the LC-VCO: (a) phase noise and (b) tuning range.

The simulation methodology as explained in the previous section is used to predict the impact of substrate noise on this VCO. The HFSS tool is used as the EM simulator. This HFSS environment is simulated from DC up to 60 GHz with a minimum solved frequency of 50 MHz and a maximum error in the S-parameters of 0.01. The HFSS simulation takes about 7.5 hours on an HP DL145 server. The EM simulations give the designer insight into how substrate noise propagates through the different reception points of the analog/RF circuit. The voltage and the current of the complete simulation model reveal how substrate noise propagates through the output of the analog/RF circuit.

6.3.2.1   Revealing the Dominant Entry Point

The dominant noise reception points can be identified by visualizing the electric fields in the HFSS environment (see Figure 6.14) [14]. Note the dark red region in the VCO core that spreads out over the entire ground plane. Substrate noise couples resistively through the bulk of the cross-coupled NMOS pair M_1 (see Figure 6.12). This can be demonstrated by connecting the bulk of the NMOS pair M_1 to the ideal zero potential ground instead of the nonideal ground plane and then resimulating the simulation model. The corresponding sideband spurs are lowered with more than 25 dB (see Figure 6.15). Thus in this case the bulk effect is the dominant coupling mechanism. This is logical since the ground plane provides a low impedance path to the off-chip ground. This is consistent with Chapter 4. There it was pointed out that for ground resistances lower than 0.65Ω the bulk effect dominates. For larger values of the resistance of the ground interconnect, ground bounce dominates.

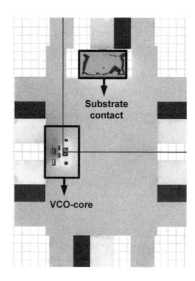

Figure 6.14   Simulated electrical fields at 100 MHz.(See color section.)

When substrate noise reaches the bulk of transistor pair M_1, it causes a voltage drop at the bulk of those transistors. This voltage drop results in modulation of the LO signal. The type of modulation is revealed by placing an ideal limiter at the output of the VCO. The limiter will remove the amplitude modulated (AM) spurs since they do not have a constant envelope. However, the limiter will not remove the spurs resulting from FM modulation because the modulated signal has a constant envelope. The spurious tones are caused by variations of the zero crossings at the output of the VCO. Figure 6.16 reveals that the spurs are FM modulated up to noise frequencies of 100 MHz. For this type of modulation the resulting sideband spurs are decreasing at a rate of 20 dB/decade with the offset frequency. Starting from 100 MHz, spurs are modulated both in amplitude and frequency (see Figure 6.16).

6.3.2.2   Modeling of the Substrate Noise Impact

At low noise frequencies, the sideband spurs are FM modulated. Chapter 5 introduced FM sensitivity functions to describe the FM spurs. An FM sensitivity function

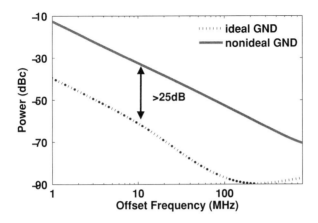

Figure 6.15 The bulk of transistor pair M_1 is the most sensitive to ground bounce. (See color section.)

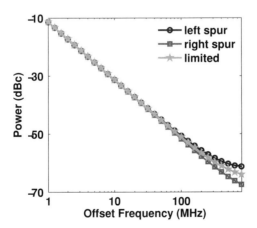

Figure 6.16 Simulated left o and right □ spur. The ★ line reflects the spurs limited at the output.

$K_i(V_{tune})$ determines how much frequency modulation results from a perturbation that is applied on a certain node of the circuit. In the previous chapter it is shown that the power of the sideband spurs can be described by [15]:

$$|V_{out}(f_{LO} \pm f_{noise})| \propto |H_{sub}^i K_i(V_{tune}) \frac{A_{LO} \cdot A_{noise}}{2 f_{noise}}| \qquad (6.3)$$

with H_{sub}^i the transfer function of the substrate contact to the bulk of the transistor pair M_1, A_{LO} the amplitude of the LO signal, A_{noise} the amplitude of the substrate noise signal at the bulk of the transistor pair M_1, and f_{noise} is the frequency of the substrate noise signal.

The simulation methodology can be used to simulate the different FM sensitivity functions and hence to determine for which values of the tuning voltage the VCO is more immune to substrate noise. Figure 6.19 shows the simulated FM sensitivity function for the external nodes of the circuit because only those sensitivity functions can afterwards be validated with measurements. One can observe that for a value of the tuning voltages of 0.2V and 0.4V, the VCO is immune to ground disturbances.

6.3.3 Verification with Measurements

To prove the usefulness of the method it is mandatory to verify the simulation results with measurements. This is the only way to check whether the simulation model incorporates the dominant coupling mechanisms. This section uses two measurement setups: the first measurement setup has a good SNR but has been restricted to 50 GHz. This measurement setup is used to measure the sideband spurs. The second measurement setup has a poor SNR but can measure up to 77 GHz. This setup is used to measure the different sensitivity functions.

6.3.3.1   Measurement of the Sideband Spurs

The LC-VCO is measured using on-wafer probes. To obtain a proper impedance termination, one buffered output of the VCO is terminated with a 50Ω load impedance, while the other is connected through a bias tee to the spectrum analyzer HP8565E. Since this spectrum analyzer can measure spectra up to 50 GHz, the oscillation frequency in this experiment was tuned below 50 GHz. A large sinusoidal signal, whose frequency is swept between 1 MHz and 800 MHz, is injected through the substrate contact (see Figure 6.17). The power generated by the RF source is set between -10 dBm and 10 dBm to guarantee sufficient SNR.

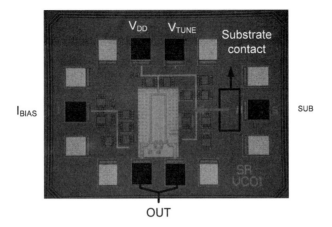

Figure 6.17   Die photograph of the 48–53 GHz LC-VCO.

Figure 6.18 shows the sideband spurs for a tuning voltage of 0.5V. The small resonances starting from a 10 MHz offset frequency are due to resonances in the different 1 mm to 2.4 mm conversion pieces and the millimeter-wave bias tee. The circuit simulations performed earlier predict an oscillation frequency that is 7% higher than the measured value. Nevertheless, the measurements and simulations of the spurs show a very good agreement both in behavior (−20 dB/decade) and in value. The careful reader will notice that starting from 100 MHz the left and right spurs do not have the same power. This indicates that those spurs in this frequency region are both modulated in frequency as in amplitude, as predicted with our simulation approach. One can conclude that our simulation approach correctly predicts the sideband spurs caused by substrate noise impact.

6.3.3.2   Measurement of the Different Sensitivity Functions

In order to measure the sensitivity functions of the different terminals of the VCO (see Figure 6.12) over the full tuning range, a single-ended output of the LC-VCO is downconverted with a dedicated millimeter-wave mixer to baseband frequencies. Then the downconverted signal is measured with a spectrum analyzer. The mixer has a conversion loss of 25 dB and hence the measurement setup has a poor SNR.

Figure 6.19 shows the measured and simulated sensitivity functions for the different terminals of the VCO. Considering the extremely high sensitivity of this VCO and taking into account that the RF models of the transistors are validated

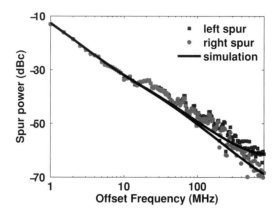

Figure 6.18   Measured versus simulated sideband spurs for a tuning voltage of 0.5V.

only up to 20 GHz, the agreement between the measurements and the simulations is acceptable.

6.3.4   Conclusions

This section uses the FDM method to characterize the substrate in order to predict the impact of substrate noise on analog/RF circuits. The methodology uses an EM simulator to characterize the substrate and the interconnects. The transistors are described by their RF models. The RF models of the transistors are cosimulated with the finite element model of the substrate and the interconnects. The methodology is successfully applied on a 48–53 GHz LC-VCO. In this particular case simulations have pointed out that substrate noise couples into the bulk of the cross-coupled transistor pair. The bulk effect dominates because this LC-VCO uses a ground plane, which offers a low impedance path to the off-chip ground.

The main drawback of the simulation methodology is the long simulation time that is required to predict the impact of substrate noise on individual analog/RF circuits. Chapter 7 relieves the simulation burden of the EM simulator and speeds up the total simulation significantly. Hence, the proposed methodology can be used to predict the impact of substrate noise on complete analog/RF systems.

Both simulation methodologies are able to provide the designer with a large amount of information about the substrate noise coupling mechanisms. This information can now be used to increase the *substrate noise immunity* of the analog/RF

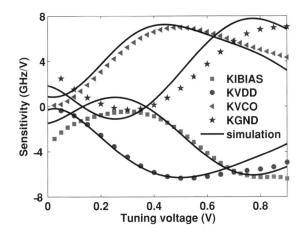

Figure 6.19   Measured versus simulated sensitivities of the different terminals of the VCO versus the tuning voltage.

circuits. The next section proposes layout and circuit techniques to improve the immunity against substrate noise. Section 6.4 explores if 3D stacking offers an opportunity to reduce the substrate noise coupling.

6.4   TECHNIQUES TO REDUCE SUBSTRATE NOISE COUPLING

The impact of substrate noise depends on three aspects:

- The generation of substrate noise;
- The transfer function from the noise source to the reception point;
- The sensitivity of the reception point to substrate noise.

The noise source can be isolated from the reception point by using guard rings. The use of guard rings was covered in Chapter 3.

The reception points of the analog/RF circuit can be desensitized at two stages of the design cycle: on the layout level and the circuit level. This section translates the insight that is gathered about the substrate noise coupling mechanisms into layout and circuit techniques, which make the VCO more immune to substrate noise.

Finally, the traditional SoC approach is questioned, and it will be shown that 3D stacking offers an opportunity to reduce the substrate noise coupling.

6.4.1 Layout Techniques to Reduce the Substrate Noise Coupling

This section present layout techniques to reduce the substrate noise coupling in LC-VCOs. First, layout guidelines are formulated to reduce the FM modulated spurs. Then, countermeasures are proposed to reduce the AM modulated spurs.

6.4.1.1 Reducing the FM-Modulated Spurs

The FM modulated spurs are mostly caused by perturbations on the ground interconnect. To reduce the power of the FM modulated spurs, the impedance of the on-chip ground interconnect has to be small compared to the impedance of the PCB ground network and the connection between the on-chip ground and the PCB ground network. In the ideal case the impedance of the ground connection is zero: no voltage fluctuation can occur at the leads of the transistors of the VCO that are connected to the ground network. Consequently, no perturbation on the ground interconnect is possible and thus no modulation of the capacitance of the LC tank. Hence, there are no spurs resulting from ground bounce.

The impedance of the ground connection can be lowered by using wider ground traces or using a thicker metal for the routing of the ground interconnect. For a same amount of current, the voltage fluctuation over the ground resistance decreases with decreasing values of ground resistance. It is interesting to look at the evolution of the sheet resistance as a function of the technology scaling. Table 6.1 summarizes the sheet resistance of the lowest metal layer for technology nodes ranging from 0.25 μm down to 45 nm.

Table 6.1

The Evolution of the Sheet Resistance of Metal 1 for Different Technology Nodes

Technology node	Sheet resistance (mΩ/square)
0.25 μm	50
0.18 μm	60
0.13 μm	70
90 nm	100
45 nm	180

In order to obtain the same impedance of the ground interconnect, the tracks in a 45 nm technology should be 3.3 times wider than in a 0.25 μm technology. This is of course contradictory with the scaling of the transistors dimensions, where more and more transistors are integrated on a smaller area. Thus, it can be concluded that the impact of substrate noise on LC-VCO worsens with technology scaling.

6.4.1.2   Reducing the AM Modulated Spurs

At intermediate frequencies substrate noise couples capacitively in the inductor of the LC-VCO and causes AM modulated spurs. The parasitic capacitance of the inductor can be lowered using a smaller footprint for the inductor. This is not an easy task since the dimensions of the inductor determine the inductance and the Q-factor of the inductor. Nevertheless, measurements show that a reduction of the footprint of the inductor moves the corner frequency where spurs change from FM toward AM modulated spurs to higher frequencies (see Figure 6.20). Of course, the position of this corner frequency not only depends on the footprint of the inductor but also on the power of the FM modulated spurs. Figure 6.20 shows the 900 MHz LC-VCO that has an inductor with a footprint of 700 μm by 700 μm. This inductor has a parasitic capacitance of approximately 2 pF. The corner frequency is around 40 MHz. The 5–7 GHz uses a smaller footprint which measures only 200μm by 200 μm. The resulting parasitic capacitance is approximately 200 fF. Figure 6.20 shows that the resulting AM-FM corner frequency is around 100 MHz. In the case of the 48–53 GHz VCO, the footprint of the inductor is only 50 μm by 150 μm. This inductor has a parasitic capacitance of only 10 fF. In this case, the parasitic capacitance of the traces are of the same order of magnitude as the parasitic capacitance of the inductor. Substrate noise may or may not couple at high frequencies into the inductor. The AM-FM corner frequency is above 800 MHz.

6.4.2   Circuit Techniques to Reduce the Substrate Noise Coupling

This section presents circuit techniques to reduce the substrate noise coupling in LC-VCOs. Circuit techniques are presented that reduce the FM and AM modulated spurs. The first two techniques focus on reducing the FM sensitivity to ground disturbances (K_{GND}). Two circuit techniques are proposed to reduce this sensitivity. The first circuit technique splits the varactors of the LC-tank of the VCO in multiple varactors. In this way it will be shown that the capacitance of the LC-tank is less sensitive to ground bounce. The next circuit technique uses an NMOS current mirror instead of a PMOS current mirror to reduce the sensitivity to ground bounce.

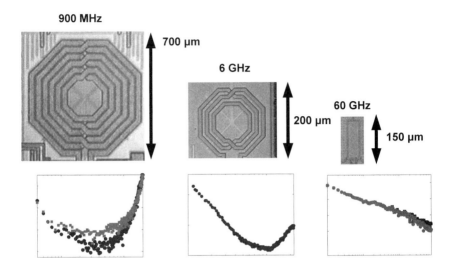

Figure 6.20   A smaller footprint moves the AM-FM corner frequency to higher frequencies.

The last circuit technique proposes a limiter to remove the AM-modulated spurs.

6.4.2.1   Switched Varactor VCO

The sensitivity of the LC-tank of the LC-VCO can strongly be reduced by using a switched varactor VCO. In such a design, the frequency tuning of the LC-VCO is performed by using multiple varactors. A small varactor is used for continuous fine tuning. The coarse frequency tuning is performed by digitally switching the varactors. A switched varactor VCO has a lower VCO sensitivity (K_{VCO}), which allows easier PLL design. This topology also benefits from a reduced sensitivity to ground bounce. This can be visualized by simulating the sensitivity of the varactor bank of the VCO in the case of a traditional LC-VCO design and in the case of a switched varactor design.

In a traditional LC-VCO design, the varactors are all connected by one terminal to the tuning voltage. In this case the varactor bank consists of eight varactors connected in parallel. Figure 6.21 shows the simulated capacitance value versus the tuning voltage for the varactor used in a traditional VCO design. It can

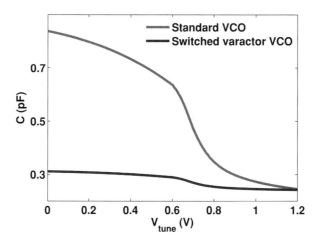

Figure 6.21   The capacitance of the varactor versus the tuning voltage.

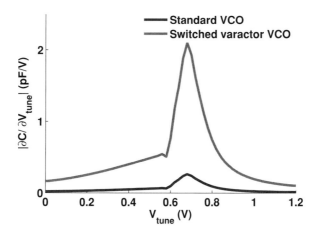

Figure 6.22   Sensitivity of the varactor bank to changes in the tuning voltage.

be seen that the capacitance value does not change much with the tuning voltage at
the edges of the tuning range. This is when the varactors are biased at the power
supply or the ground. The derivative $\partial C / \partial V_{tune}$ is representative for the sensitivity
of the LC tank of the VCO (see Figure 6.22). For a traditional LC-VCO design,
the VCO is particularly sensitive for tuning voltages around 0.6V. Even small levels
of ground bounce are sufficient to significantly change the capacitance value of the
varactors and hence the oscillation frequency of the LC-VCO.

To demonstrate the reduced sensitivity in the case of a switched varactor
VCO, seven out of eight varactors are connected to the power supply. Figure 6.22
shows the simulated capacitance value versus the tuning voltage for the different
settings of the digitally switched varactors. In this case, the curves are much flatter.
The first derivative shows that the sensitivity of the VCO is strongly reduced with
this topology (see Figure 6.22). The sensitivity can be reduced further using a
smaller varactor for the continuous fine tuning. In that case more digitally switched
varactors are needed in order to cover the whole tuning range. This makes the digital
control unit of the varactor bank more complex.

The analog designer probably wonders how small the varactor for the continuous fine tuning should be. This depends on the oscillation frequency of the VCO.
A VCO with a high oscillation frequency is much more sensitive to ground bounce
than a VCO with a low oscillation frequency. This is demonstrated by comparing the
sensitivity to ground bounce of the 900 MHz and 50 GHz VCO (see Figure 6.23).
The 50 GHz VCO is approximately 25 times more sensitive than the 900 MHz
VCO. In order to obtain the same sensitivity to ground bounce for both VCOs, the
contribution of the varactor bank that is required for fine tuning needs to be very
small. Assume that ground bounce causes 1% variation in the capacitance value of
the LC tank of the VCO. In the case of the 900 MHz VCO this causes a shift in
oscillation frequency of 5 MHz, while in the case of the 50 GHz VCO this causes a
shift of 254 MHz. The 50 GHz VCO will thus need a very small varactor to obtain
the same sensitivity as the 900 MHz VCO. This means that the digitally controlled
part of the varactor bank needs to be encoded with a high resolution in order to
cover the whole frequency range.

6.4.2.2   Using an NMOS Current Mirror

The current mirror of the LC-VCO is often a PMOS current mirror. This topology
is popular for its low phase noise potential. The PMOS devices offer lower flicker
noise than the NMOS devices [16]. In order to increase the immunity to ground
bounce, one might use an NMOS current source. An LC-VCO with an NMOS

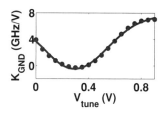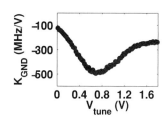

Figure 6.23   The 50 GHz VCO is approximately 25 times more sensitive than the 900 MHz LC-VCO.

current source is approximately 6 dB less susceptible to ground bounce than its PMOS counterpart. However, this will make the power supply lines more sensitive to any perturbation.

6.4.2.3   Limiter

Placing a limiter at the output of the VCO strongly suppresses the AM modulated spurs. It might not be necessary to design a dedicated limiter because the VCO is often used in front of a switching mixer. Such a switching mixer also behaves as a limiting circuit. If the AM modulated spurs are not sufficiently suppressed, one might consider the design of a dedicated limiter. The limiter, however, should be immune to substrate noise and not convert AM signals into FM signals. Putting an extra limiter between the VCO and the mixer will increase the overall power consumption of the receiver.

6.4.3   3D Stacking as a Solution to Substrate Noise Issues

Today's mobile phones feature a host of ICs ranging from analog/RF circuits, memory, digital signal processors, and so forth. New devices include image displays, MP3 devices, and image sensors. Given the large number of ICs, placing them on printed-circuit boards (PCBs) and routing them to be properly interconnected in a tight layout is a daunting task. Integrating all the functionality on a single die is at this moment not possible. To further reduce the cost factor of mobile phones, the semiconductor industry tends to move to 3D stacked ICs. Such an IC is also known as a system in a cube (see Figure 6.24). A 3D SoC is a chip with more layers of active electronic components, integrated both vertically and horizontally into a single system. The semiconductor industry is hotly pursuing this promising technology in many ways. One way to stack 3D ICs is the *die-on-die* method [17, 18]. The electronic circuitries are built on two or more dies, which are then stacked on top of each other. The different layers are connected with each other through vertical connections, called *through-silicon vias* (TSVs). This technology offers many significant benefits. First of all, 3D SoCs save on IC footprint area. Furthermore, thanks to the TSVs, the signals do not have to go off-chip anymore, which greatly reduces the power consumption and the propagation delay and extends the usable signal bandwidth.

Also the cost can be greatly reduced because in comparison with the SoC, the electronics circuitries do not need to be integrated in the same technology. For example the power management unit can be fabricated in a cheap 1 μm technology and the fast digital processor can be designed in an expensive 45 nm technology.

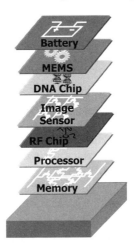

Figure 6.24   3D stacking increases the immunity against substrate noise.

The next section explores, with simulations, the benefits of 3D stacking for reducing the impact of substrate noise on an analog/RF circuit and the validation of those simulations with measurements. Therefore, we consider a 3D SoC consisting of two dies. The bottom die contains the aggressor, which is a substrate contact that serves to emulate the switching activity of the digital circuitry [15]. The top die contains a 48–53 GHz millimeter-wave LC-VCO that acts as the victim. The aggressor on the bottom die and the victim VCO on the top die couple mainly in two ways with each other:

- Capacitive coupling occurs from the bottom die to the top die;

- Substrate noise causes ground bounce on the bottom die. This ground bounce is sensed by the nonideal off-chip ground and couples through this off-chip ground back into the ground interconnect of the VCO located on the top die.

The impedance of the ground connection determines the importance of both coupling mechanisms. Both coupling mechanisms are quantified with measurements on a real-life silicon prototype.

6.4.3.1   Description of the 3D SoC Experiment

The 3D SoC experiment is set up based on the die-on-die method of stacking. This method consists of building the different circuitries that are possibly realized in different technologies, on two or more dies. In our case we consider the stacking of two identical dies without using any TSVs. Both dies consists of a substrate contact and a millimeter-wave VCO. In this way we can also simulate and measure the substrate isolation in the case of a 2D SoC. This allows a fair comparison of the substrate noise isolation between a 2D SoC and a 3D SoC.

The bottom die in the case of the 3D SoC acts as the aggressor. The VCO of the bottom die is thus not active. In this experiment a substrate contact is used to emulate the switching activity of the digital circuitry (see Figure 6.25).

The top die consists of the 48–53 GHz LC-VCO that is already used in this chapter (see Figure 6.25). The substrate of the top die is thinned to 25 μm (see Figure 6.25). Usually the thinning varies from 25 μm up to 50 μm. Thus, our case considers the worst case where the bottom circuitry is the closest to the top circuitry. Further, the substrate contact located on the top die is removed by a dicing operation (see Figure 6.25). This allows us, during measurements, to reach the bondpad connected to the substrate contact of the bottom die with on-wafer probes. Then, the die is glued with a polymer material called BCB on the bottom die (see Figure 6.25). The polymer layer is a nonconductive layer with a thickness of approximately 1 μm and a ϵ_r of 3.

6.4.3.2   Impact Simulation Methodology

A similar simulation methodology is used as described in Section 6.3. However, this methodology is extended to 3D SoCs. The simulation approach consists of two simulations. First, an EM simulation [2] is performed. This simulation includes the on-chip interconnects, the substrate and the passive components. Next, the results of

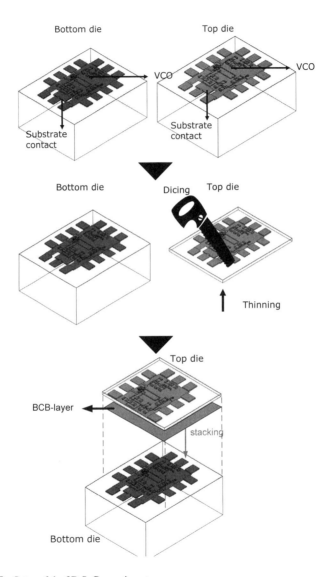

Figure 6.25   Setup of the 3D SoC experiment.

the EM simulation are used together with the RF models of the devices to perform a SpectreRF [12] simulation. The resulting waveforms on the different terminals of the simulation model will give the designer insight in the different substrate noise coupling mechanisms. This section discusses the different simulations that are needed to characterize the 3D SoC experiment:

- *EM simulation:* The simulation setup is similar to Section 6.3. The layouts of the bottom and top circuitries are streamed into the EM environment. In this environment the substrate, the silicon dioxide, and the p-well are added for each die (see Figure 6.26). An air box is added on top of the top die and a BCB layer connects both dies with each other. Furthermore, one port is placed at the substrate contact of the bottom die. On the top die, ports are placed at the connections of the different bias lines, at the terminals of the transistors and the differential output of the VCO. A complete cross section of the experiment is shown in Figure 6.26.

  This EM model is simulated from DC up to 60 GHz with a minimum solved frequency of 50 MHz and a maximum ΔS of 0.01. It takes 7.5 hours on an HP DL145 server to simulate the S-parameters at the different ports.

- *Circuit simulation:* A simulation model is constructed that fully characterizes the 3D-IC. The interconnects, the substrate, and the passive components of both dies are represented by an S-parameter box resulting from the EM simulation. The transistors are represented by their RF model and are properly connected to the S-parameter box. On this complete simulation model the designer can apply any circuit analysis. The goal of this experiment is to predict the power of the substrate noise induced sideband spurs. To that end, a periodic AC analysis is performed on this simulation model with SpectreRF. The analysis takes 2 minutes on a HPUX9000 platform.

6.4.3.3   Reducing the Impact of Substrate Noise

An LC-VCO is very sensitive to substrate noise coupling [15, 19]. A substrate noise signal, which couples into a VCO, will modulate the oscillator signal both in frequency and in amplitude. These modulation effects cause sideband spurs to appear around the local oscillator. In this section the power of the sideband spurs in a 3D SoC design is compared to the traditional 2D SoC solution.

In [20] where the impact of substrate noise was predicted on the same VCO in a 2D SoC context, all the ground bond pads were shorted to a zero potential ground. This makes sense since all the ground bond pads do share the same ground

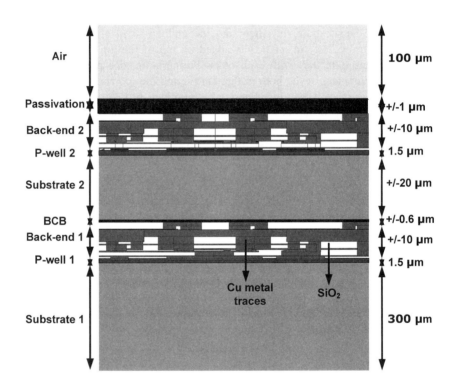

Figure 6.26 Cross section of the 3D stacked experiment.

plane. This is not the case when the 3D SoC is considered. The amount of gained isolation, however, strongly depends on the grounding scheme. We consider two different grounding schemes (see Figure 6.27):

- The digital/analog grounds (bottom/top die) are perfectly separated. This is the ideal case because both grounds need to be connected anyway. The connection is usually made on the PCB.

- The digital/analog grounds are connected together on the PCB and there exists a resistance between the point where both grounds meet each other and the ideal zero potential ground. In a real scenario, this is the earth of the building.

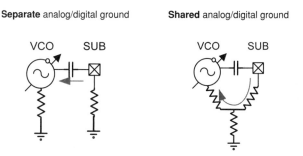

Figure 6.27   The substrate noise isolation is determined by the impedance of the ground connection.

Substrate noise is injected into the substrate contact of the bottom die while the power of the spurs that appear at the output of the VCO, located on the top die, is sensed. The power of the spurs versus the frequency distance to the local oscillator of the VCO is simulated for both the 3D SoC and the 2D SoC case and also for the two different grounding schemes. The next section discusses the different simulations.

6.4.4   Separated Analog/Digital Ground

In the first case both dies are connected separately to the ideal ground. This means that the impedance between the node where the digital and analog ground meet each other and earth (i.e., ideal ground), is zero. This is the ideal case. Substrate noise is injected into the substrate contact of the bottom die. From there on, it propagates through the substrate of the bottom die and then couples through the silicon dioxide and the BCB layer capacitively to the top die (see Figure 6.28). In the 2D SoC case, the spurs are decreasing at a rate of 20 dB/decade because the spurs are mainly modulated in frequency [20]. Due to the capacitive coupling the spurs are not decreasing at a rate of 20 dB/decade anymore in the 3D SoC case. Due to capacitive coupling the power of the spurs almost does not vary with the frequency of the injected signal.

Compared to the 2D SoC solution, 3D stacking improves the immunity with 100 dB for a substrate noise signal of 1 MHz. For a noise signal of 800 MHz the immunity improves with 40 dB (see Figure 6.28). A huge amount of isolation can be achieved because the capacitance between the two dies can be made as small as 70 fF.

6.4.5   Shared Analog/Digital Ground

In the case where both the analog and digital ground are connected to each other, a small resistor of only 0.5Ω is placed between the node where both ground connections meet each other and earth (see Figure 6.29). In this case, less substrate noise isolation is obtained (see Figure 6.29). Compared to the traditional SoC solution the substrate noise isolation improves only 20 dB for a noise signal of 100 MHz. For a noise signal of 800 MHz the improvement is only 10 dB. Moreover, no change is observed in the behavior of the spurs. In both the 2D SoC and the 3D SoC, the spurs are decreased by 20 dB/decade.

In the case of a shared analog/digital ground, substrate noise is injected into the substrate contact of the bottom die. This causes ground bounce on the bottom die. This ground bounce is sensed by the PCB ground and causes a voltage fluctuation over the 0.5Ω resistance. The ground bounce on the PCB ground is directly sensed by the VCO ground on the top die. Ground bounce on the VCO ground also causes FM modulated spurs, which decrease by 20 dB/decade (see Figure 6.29).

6.4.6   Experimental Validation

The only way to determine which of the coupling mechanisms is the most important in a practical context is to perform measurements on a real-life silicon prototype. To that end, a 3D stacked experiment is set up as described in Section 6.4.3.1. Figure 6.30(a) shows the top view of the 3D stacked experiment. Note the almost perfect alignment of the two dies. The black horizontal line on the photograph reflects the transition region from the top die to the bottom die. Figure 6.30(b) is

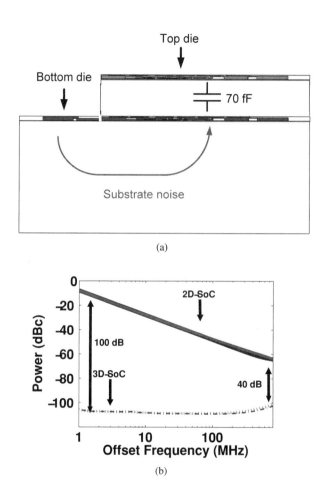

(a)

(b)

Figure 6.28   First case. (a) Simulation results in the case of an ideal PCB ground. (b) Substrate noise couples capacitively to the VCO.

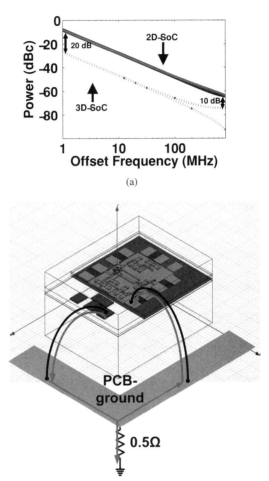

(a)

(b)

Figure 6.29 Second case. (a) Simulation results in the case of a shared analog/digital ground. (b) Substrate noise couples back from the PCB toward the VCO.

a 3D picture that zooms in on this transition region. The height difference between the top and the bottom die is 27 μm.

(a)

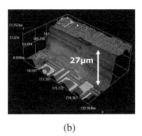
(b)

Figure 6.30   Picture of the 3D stacked experiment (a) Top view and (b) 3D image of the stack.

The 3D stacked experiment is measured using on-wafer probes. To obtain a proper impedance termination, one buffered output of the VCO is terminated with a 50Ω load impedance, while the other is connected through a bias tee to the spectrum analyzer HP8565E. Since this spectrum analyzer can measure spectra up to 50 GHz, the oscillation frequency in this experiment was tuned below 50 GHz. A large sinusoidal signal, whose frequency is swept between 1 MHz and 800 MHz, is injected through the substrate contact. The power generated by the RF source is set between -10 dBm and 10 dBm to guarantee sufficient SNR. The measurement noise floor is -65 dBc.

The performance of the VCO in the 3D SoC case is comparable with the 2D SoC case. Only 1% of deviation in the oscillation frequency is observed. Hence, one can conclude that the thinning and the stacking does not have a visible effect on the performance of the stacked VCO. Figure 6.31 compares the measured spurs both in the case of a 2D SoC and a 3D SoC. At low frequencies a 3D SoC improves the substrate isolation by 20 dB when compared with the 2D SoC approach. However at frequencies of 300 MHz, there is no difference in the power of the spurs between a 2D SoC and a 3D SoC.

During measurements, a high sensitivity of the power of the spurs to the impedance of the ground connection was observed. Together with the fact that at low frequencies the spurs measured on the 3D SoC are decreasing at a rate of 20 dB/decade, we are confident that substrate noise that is injected into the substrate contact of the bottom die causes ground bounce on the measurement ground. This ground bounce couples back to the ground of the VCO on the top die and causes spurs in the output spectrum of the VCO.

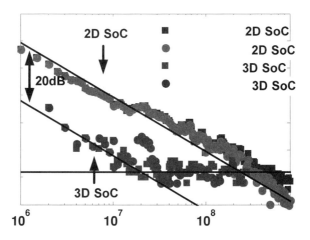

Figure 6.31   Measured spurs for the 3D SoC versus the 2D SoC.

The measured resistance from the measurement ground (i.e., ground of the wafer probe station) to earth is 0.5Ω. Therefore one can conclude that at low frequencies there is a good agreement between measurements and simulations. However at frequencies as low as 30 MHz, there is a discrepancy between measurements and simulation. From 30 MHz on, the spurs are frequency independent. This is probably due to the inductive nature of the ground connection. Because the impedance of the ground connection has a distributed nature and could not be directly measured, it could not be included in the simulations.

6.4.7   Conclusions

In this section we have shown that the knowledge about substrate noise coupling mechanisms that is provided by the different simulation methodologies can be translated into different layout and circuit techniques that reduce the substrate noise coupling. The proposed layout techniques consist of widening the traces of the ground interconnect to reduce the FM modulated spurs and shrinking the footprint of the inductor to reduce the AM modulated spurs. Three circuit techniques are proposed to reduce the substrate noise coupling:

- Use a switched varactor VCO to reduce the sensitivity of the VCO to ground bounce.

- Use an NMOS current source instead of a PMOS current source in the design of the LC-VCO.

- Place a limiter at the output of the VCO to remove the AM modulated spurs.

Further, this section shows that 3D stacking has a potential to improve the substrate noise immunity of analog/RF circuits. However special attention should be paid to the PCB grounding scheme. The ground interconnects of the different wafers should be connected on the PCB as close as possible to the *ideal* ground.

6.5 CONCLUSIONS

Simulations are shown to be adequate to predict the impact of substrate noise in analog/RF circuits with a good accuracy when compared to measurements. Moreover, simulations are able to retrieve the nature of the different coupling mechanisms. This chapter proposes two simulation methodologies:

- The first methodology models the substrate with the FDM method. This methodology enables one to reveal the different substrate noise coupling mechanisms for a 900 MHz LC-VCO. At low frequencies (<10 MHz) substrate noise couples into the nonideal ground of the VCO and causes ground bounce. The resulting spurs are FM modulated. At higher frequencies (>100 MHz) substrate noise couples capacitively into the inductor of the LC-VCO and causes AM modulated spurs.

- The second methodology models the substrate with the FEM method. This methodology is applied to a challenging example, which is a 48–53 GHz millimeter-wave VCO. Here, substrate noise couples dominantly into the bulk of the cross-coupled pair of the VCO.

Both methodologies conclude that at low noise frequencies, substrate noise usually results in ground-related problems.

The information about the different substrate noise coupling mechanisms that is obtained by the two methodologies is translated into circuit and layout techniques, which reduce the sensitivity of the VCO to substrate noise. Hence, the methodologies prove to be very useful during the design phase and open the way toward a substrate noise aware design. The second methodology is extended toward

3D stacked ICs. Simulations show that 3D stacking offers a potential to reduce the substrate noise coupling when compared to the 2D SoC solution.

This chapter compares the performance of both methodologies and argues that the second methodology is the preferred methodology to study the impact of substrate noise on *analog/RF systems* because it circumvents most of the shortcomings of the first methodology. The next chapter refines the second methodology with the a priori knowledge that substrate noise coupling in analog/RF circuits usually results in ground-related problems and applies this refined methodology on a small receiver.

References

[1] Substrate Noise Analysis Cadence, http://www.cadence.com.

[2] HFSS, http://www.ansoft.com/products/hf/hfss/.

[3] DIVA, http://www.cadence.com/products/dfm/divaS.

[4] SNA, Substrate Noise Analysis Cadence, http://www.cadence.com.

[5] J. Schoukens, R. Pintelon, and T. Dobrowiecki, "Linear modeling in the presence of nonlinear distortions," *IEEE Transactions IM*, Vol. 51, No. 4, August 2002, pp. 786–792.

[6] *Frequency Domain Identification Toolbox, V3.3 for MATLAB.*, Gamax Ltd, Budapest, 2005.

[7] C. Paul, "Effectiveness of multiple decoupling capacitors," *EMC*, Vol. 34, No. 2, May 1992, pp. 130–133.

[8] C. Soens, *Modeling of Substrate Noise Impact on CMOS VCOs on a Lightly Doped Substrate*,   Acco, 2005.

[9] C. Soens, G. Van der Plas, P. Wambacq, and S. Donnay, "Substrate noise immune design of an LC-tank VCO using sensitivity functions," *Proc. Custom Integrated Circuits Conference the IEEE 2005*, 2005, pp. 477–480.

[10] X. Lai and J. Roychowdhury, "Capturing oscillator injection locking via nonlinear phase-domain macromodels," *IEEE Transactions MTT*, Vol. 52, No. 9, 2004, pp. 2251–2261.

[11] http://eesof.tm.agilent.com/products/rfde2003c_momentum.html.

[12] SpectreRF, http://www.cadence.com/products/custom_ic/spectrerf.

[13] B. Razavi, "A 60-GHz CMOS receiver front-end," *IEEE Journal of Solid State Circuits*, Vol. 41, No. 1, January 2006, pp. 17–22.

[14] D. White and M. Stowell, "Full-wave simulation of electromagnetic coupling effects in RF and mixed-signal ICs using a time-domain finite-element method," *IEEE Transactions MTT*, Vol. 52, No. 5, 2004, pp. 1404–1413.

[15] C. Soens, G. Van der Plas, P. Wambacq, S. Donnay, and M. Kuijk, "Performance degradation of LC-tank VCOs by impact of digital switching noise in lightly doped substrates," *IEEE Journal of Solid State Circuits*, Vol. 40, No. 7, July 2005, pp. 1472–1481.

[16] Y. Tsividis, *Operation and Modeling of the MOS Transistor*, Oxford, U.K.: Oxford University Press, 1999.

[17] E. Beyne, "3D system integration technologies," *Proc. International Symposium on VLSI Technology, Systems, and Applications*, April 2006, pp. 1–9.

[18] P. Pieters and E. Beyne, "3D wafer level packaging approach towards cost effective low loss high density 3D stacking," *Proc. 7th International Conference on Electronic Packaging Technology ICEPT '06*, 26–29 August 2006, pp. 1–4.

[19] S. Bronckers, C. Soens, G. Van der Plas, G. Vandersteen, and Y. Rolain, "Simulation methodology and experimental verification for the analysis of substrate noise on LC-VCO's," *Proc. Design, Automation & Test in Europe Conference & Exhibition, DATE '07*, 2007, pp. 1–6.

[20] S. Bronckers, K. Scheir, G. Van der Plas, and Y. Rolain, "The impact of substrate noise on a 48-53GHz mm-wave LC-VCO," *Proc. IEEE Topical Meeting on Silicon Monolithic Integrated Circuits in RF Systems SiRF '09*, 2009, pp. 1–4.

Chapter 7

Noise Coupling in Analog/RF Systems

7.1 INTRODUCTION

Chapter 6 proposed a methodology to predict the impact of substrate noise in analog/RF circuits. This methodology combines the strengths of the EM simulator and the circuit simulator. The passive part, which consists of the substrate and the interconnects is simulated with an EM solver. The resulting S-parameters are elegantly combined with the RF models of the active devices in one simulation model. The impact of substrate noise is predicted by an evaluation of this simulation model with a circuit simulator. The simulation methodology accurately predicts the impact of substrate noise on analog/RF circuits. However, the long simulation time hampers the methodology in being applied on practical analog/RF systems. The bottleneck is formed by the simulation burden of the EM solver. The EM simulator needs to solve all the small details of the interconnects, which is a very CPU-intensive process. This chapter relieves the simulation burden of the EM simulator and significantly reduces the total simulation time. This is possible by taking advantage of the a priori knowledge. At low frequencies substrate noise couples into the ground interconnect of the analog circuitry [1–5]. Hence, the EM simulator only needs to solve the substrate together with the ground interconnect. The other interconnects can easily be modeled by a parasitic extraction and included as an RC mesh in the simulation model. In this way the simulation time can be significantly reduced. Hence, the improved version of the methodology is capable of predicting the impact of substrate noise in analog/RF systems with a good accuracy and a reasonable simulation time.

This chapter leads the reader through the complete simulation and extraction process of how to correctly set up the simulation environment to predict the impact of substrate noise on analog/RF systems. Further, the performance of the improved simulation methodology is compared to the performance of the simulation methodology of Chapter 6. Both methodologies are used to predict the impact of substrate noise on the same analog/RF circuit. The improved simulation methodology shows to be at least an order of magnitude faster without compromising the accuracy. Finally, it will be shown that the improved simulation methodology is able to predict the impact of substrate noise in complex analog/RF systems with a good accuracy and a reasonable simulation time.

7.2   IMPACT SIMULATION METHODOLOGY

The simulation methodology in Chapter 6 had one major drawback, which is the long simulation time of the EM solver. Indeed, the EM simulator needs to simulate all the geometric details of the interconnects and the substrate, which is very CPU-intensive. Therefore, the EM simulator needs a lot of CPU power and several hours to characterize all the interconnects and the substrate. Due to this, the methodology is limited to (small) analog/RF circuits. In order to predict the impact of substrate noise on large analog/RF circuits and even on complete analog/RF systems, the simulation burden of the EM simulator needs to be reduced.

The simulation time of the EM simulator can only be reduced if approximations are made. To obtain good quality simulations using a streamlined approximation process, one has to take advantage of a priori knowledge. Here, we know that substrate noise couples at low frequencies into the ground interconnect of the analog circuitry. The frequency range where this is valid depends of course on the analog/RF system under test. As a rule of thumb, one can assume that for noise frequencies lower than 100 MHz, substrate noise couples into the ground interconnect of the analog/RF circuit. Hence, the analog circuit either suffers from the bulk effect or ground bounce. In order to describe the propagation of substrate noise from the noise source to the different reception points of the analog/RF system, only the substrate with its different doped regions and the ground interconnect need to be taken into account by the EM simulator. Hence, the EM simulator does not need to simulate all the small details of the other interconnects. This significantly simplifies the EM simulation test bench and reduces the EM simulation time. Hence, the substrate and the ground interconnect are characterized by an S-parameter model.

Of course, there is still a need for a model of the signal interconnects to obtain a full simulation model of the circuit. Here, the idea is to trade the full EM model for a lumped equivalent circuit of the connections. Fortunately, a commercially available tool can be used to model the signal interconnects. A parasitic extraction models the signal interconnects as an RC-mesh. This model of the interconnects can easily be integrated into the complete simulation model. Note that the simulation model now uses three types of models: EM, RC meshes, and transistor models.

The RF models of the active devices are elegantly combined with the model of the interconnects, the model of the substrate, and the ground interconnect in one simulation model. This simulation model characterizes the entire analog/RF system. This simulation model can be used by the designer to apply any circuit analysis.

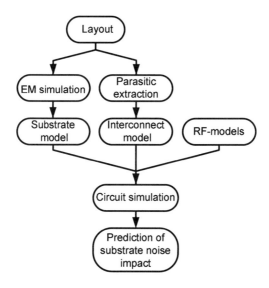

Figure 7.1 Block diagram of the proposed methodology.

A block diagram representation of the methodology is given in Figure 7.1. The different simulations that are needed to predict the impact of substrate noise on analog/RF systems and the user interaction that is required are described in more detail.

7.2.1   EM Simulation

The goal of the EM simulation is to characterize the propagation of substrate noise from the noise source towards the different reception points of the analog/RF system. Similar to the simulation methodology used in the previous chapter, the EM simulation starts from the layout of the analog/RF circuit. The conversion from the layout to the EM environment involves four steps:

- Simplifying the layout;

- Building the EM environment;

- Setting up the EM simulation;

- Approximating the substrate transfer function.

    Each step requires a certain amount of user interaction as will be explained below in detail.

7.2.1.1   Simplifying the Layout

The layout is simplified to reduce the simulation burden of the EM simulator. In contrast with the simplifications of Chapter 6, this simulation methodology no longer characterizes all the interconnects. The layout is stripped down until the ground interconnects and their connections to the substrate remain. The remaining geometry can be further simplified by applying the same simplifications as used in the simulation methodology of the previous chapter. The simplifications are summarized below (refer to Chapter 6 for more details on how to perform these simplifications):

- The different vias that connect the different metal layers are grouped.

- The edges of the interconnects are aligned.

- The ground plane that is slotted to satisfy the DRC rules is replaced by a filled metal layer.

- The transistors are removed from the layout as the EM simulator is not able to solve the drift diffusion equations, which are needed to characterize the behavior of the transistors. As in Chapter 6, the idea is to place ports at the remaining transistor leads.

7.2.1.2   Building the EM Environment

The simplified layout is then streamed into the EM environment. The substrate, the p-well, the silicon dioxide and the air box on top of the circuit are included in this environment. Silicon boxes with zero conductivity are inserted around the n-wells to model the depletion region of the corresponding PN junction.

Then ports are placed. These ports form the interface between the EM model and the external world. In this simulation methodology each transistor can be described with only one port. It is instructive to illustrate the reduction of the total number of ports with a small example. Consider a common source transistor. In the previous chapter all the interconnects were modeled in the EM environment. To characterize this transistor three ports were placed. The drain, bulk, and gate of the transistor were referred to its source. In the methodology used here, the gate and drain connections of the common source transistor are modeled by a parasitic extraction tool and therefore no ports need to be placed at those connections. Only one port is placed at the bulk of the transistor referred to its source.

A circuit consisting of n transistors and one substrate contact therefore needs only $n + 1$ ports. Consequently the EM simulator needs to solve $(n + 1)^2$ S-parameters. Compared to the simulation methodology of the previous chapter, this is a significant reduction of the total number of ports. The simulation methodology of the previous chapter needs $3n$ ports per transistor and thus the EM simulator solves $(3n + 1)^2$ S-parameters. Reducing the number of ports lowers the complexity of the simulation and also prohibits convergence problems.

Finally, ports are placed at the external ground connections.

7.2.1.3   Setting Up the EM Simulation

Once the EM environment is built, the EM simulation can be set up. The S-parameters are solved at the different ports for a user-given frequency range and tolerance. The resulting S-parameters characterize the substrate and the ground interconnect.

7.2.1.4   Approximating the Substrate Transfer Function

This last step is not always necessary. This step guarantees smooth convergence of the circuit analysis and ensures good extrapolation to DC of the S-parameters. Therefore, a lumped model is built that approximates the transfer function from the

aggressor (the substrate contact) to the different reception points of the analog circuit (the victim). The resulting S-parameter matrix is converted into a Y-parameter
matrix. In this Y-parameter matrix Y_{ij} reflects the admittance from node i to node
j. The different Y_{ij} values are approximated by a lumped network. Since the EM
environment consists of a large number of ports, this may require a lot of modeling.
However, one can define a global admittance function for a set of ports that share
the same ground plane. The admittance Y_{ij} between the nodes that share the same
ground plane is close to infinity (resistance close to zero) because the different nodes
are almost shorted by the ground plane. Hence, the global admittance function from
the aggressor (the substrate contact) to the set of nodes that share the same ground
plane can be approximated as

$$Y_{global} = \sum_i Y_{i,j} \qquad (7.1)$$

where i reflects the nodes that share the same ground plane, and j reflects the aggressor node. Since the aggressor is in our case a substrate contact, substrate noise
origins from one single node. This global admittance function is then approximated
by a lumped network. Working with global admittance functions reduces the modeling effort.

7.2.2 Parasitic Extraction

The parasitic extraction also starts from the layout (see Figure 7.1). The interconnects are meshed into rectangles. Each of the rectangles is then modeled as an RC
network. The capacitance of the interconnect is determined based on its geometry
and the dielectric properties of the oxide. The parasitic resistance of the interconnect is determined based on its geometry and its sheet resistance. The different
RC networks are then automatically connected to each other and to the active and
passive devices. Such a parasitic extraction can be performed with software like
Calibre [6] PEX. The resulting RC network can be simplified by using an order
reduction modeling technique.

7.2.3 Circuit Simulation

A simulation model that fully characterizes the analog/RF circuit is constructed.
The three models, the substrate, and the ground interconnect are represented by an
S-parameter box resulting from the EM simulation or are represented by an approx-
imated lumped network. The model for the substrate and the ground interconnect

is included into the netlist together with the RF models of the different devices and the lumped model of the parasitics of the signal routing provided by the parasitic extraction. The analog designer can apply any analysis to this simulation model. If the system under test is differential, the mismatch between two identical transistors must be considered as well. Mismatch is caused by the process variability on the width (w) and threshold voltage (v_{th}) between two identical transistors. The statistic nature of these variations implies that it is not possible to include the exact amount of mismatch [7]. Rules of thumb were used to estimate the mismatch, namely:

$$\sigma_{vth} = \frac{5\,mV\,\mu m}{\sqrt{W \cdot L}} \quad \text{and} \quad \frac{\sigma_\beta}{\beta} = \frac{0.02\,\mu m}{\sqrt{W \cdot L}} \tag{7.2}$$

where W and L are, respectively, the width and the length of the MOS transistor, σ is the standard deviation, and β is the current factor [7]. Equation (7.2) shows that mismatch increases when the transistor scales down. To include mismatch effects the width of one of the two identical transistors is changed. Furthermore, an ideal DC voltage source is placed at the gate of this transistor to model the mismatch on the threshold voltage of the transistor.

The corresponding waveforms of the analysis will give insight into how substrate noise couples into the ground interconnect of the circuit and how the resulting ground bounce affects the performance of the circuit. How this analysis is performed is the subject of the next section.

7.3 ANALYZING THE IMPACT OF SUBSTRATE NOISE IN ANALOG/RF SYSTEMS

Once the simulations are performed, the designer can start to analyze the impact of substrate noise in the analog/RF system under test. Remember that the impact of substrate noise is determined by three aspects:

- The generation of substrate noise;
- The noise propagation mechanism, which is described by the transfer function from the noise source to the reception point;
- The sensitivity of the reception point to substrate noise.

The product of these three aspects determines the impact of substrate noise in an analog/RF system. The first aspect is not covered in this book. The digital circuitry is replaced by a substrate contact that is excited with an external signal.

Hence, the analysis of the impact of substrate noise in analog/RF systems focuses on the latter two aspects. This section explains how the different performed simulations give insight into the propagation of substrate noise and the different substrate noise coupling mechanisms.

7.3.1 Analysis of the Propagation of Substrate Noise

Substrate noise propagation is mainly an electric effect. Hence, the propagation from the substrate contact toward the different reception points of the analog/RF circuit is proportional to the electric field strength. Therefore, the substrate noise propagation can be visualized by plotting the simulated electric fields in the HFSS environment. Analyzing the electrical fields values in the whole structure determines how much substrate noise couples into a reception point.

7.3.2 Analyzing the Substrate Noise Coupling

The complete simulation model allows the designer to apply any circuit analysis. Combining different types of circuit analysis allows one to reveal how substrate noise couples through the analog/RF system. Moreover, since the designer has a simulation model to characterize the whole analog/RF system, he or she can start to manipulate the simulation model to gain insight into the substrate noise coupling mechanisms, for example:

- *Sensing the voltages and currents at the different nodes:* This allows one to monitor the propagation of substrate noise through the analog/RF circuit.

- *Disabling coupling mechanisms:* If the designer suspects that the dominant substrate noise coupling mechanism occurs through the ground interconnect of a particular transistor, he or she can connect the ground of this transistor to the ideal ground and monitor if the power of the spurs is decreased. If the power of the spurs is decreased, the designer has identified the transistor that needs to be shielded the most against substrate noise. In this way the substrate noise immunity of the whole analog/RF system can be improved iteratively.

- *Removing mismatch:* If the designer suspects that the mismatch of a particular transistor contributes to the generation of a certain spur, he can remove the mismatch on that transistor and again monitor if the power of the spurs is decreased.

Of course, there are more possible manipulations of the simulation model that give insight into the substrate noise coupling mechanisms. The type of manipulation depends on the analog/RF circuit itself and on the experience of the designer.

7.4 SUBSTRATE NOISE IMPACT ON A 48–53 GHZ LC-VCO

The proposed methodology is first applied on a small analog circuit. The reason for this is threefold. First, it allows one to show the proposed methodology under practical constraints. Second, this circuit is small enough to allow one to verify if the prior knowledge that the substrate noise coupling at low frequencies happens mainly in the ground interconnect of the circuit is applicable. Third, it also allows one to determine if this approximation does not compromise the accuracy of the power of the predicted spurs. To that end, the proposed methodology is applied on the same analog circuit as the one used in the previous section. This allows us also to assess the performance of the proposed methodology to the simulation methodology of Chapter 6.

7.4.1 Description of the LC-VCO

The VCO under test is described in Chapter 6. The VCO performs well and therefore is a good candidate to study the impact of substrate noise. This section briefly discusses the most important properties of this VCO. The VCO under test is a 48–53 GHz LC-VCO that is designed in a UMC 0.13 μm CMOS technology. The frequency of the VCO is determined by the resonance frequency of the LC tank. The losses of this LC tank are compensated by the gain of the cross-coupled NMOS transistor pair. The current through the core of the VCO is determined by an NMOS current mirror (see Figure 7.2). The VCO is buffered with source followers.

The LC-VCO offers a phase noise performance of -84 dBc/Hz at 1 MHz offset frequency and a tuning range of 10%. The VCO core draws 4.5 mA from a 1V supply.

7.4.2 Simulation Setup

The proposed methodology consists of three simulations:

1. An EM simulation that characterizes the substrate;

2. A parasitic extraction that models the parasitics of the interconnects;

3. A circuit simulation that fully characterizes the impact of substrate noise.

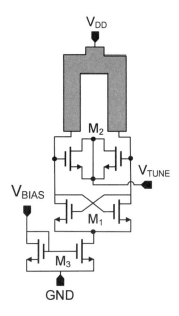

Figure 7.2   Schematic of the millimeter-wave LC-VCO.

7.4.2.1   EM Simulation

A finite element model of the substrate and the ground interconnect of the millimeter-wave VCO is built. The layout of the VCO is stripped until only the ground interconnect and the connections to the substrate remain (see Figure 7.3). This means that all the other interconnects are deleted in the layout editor. In this case the ground interconnect is laid out as a ground plane. This ground plane shields the long interconnects from the substrate. In order to demonstrate this shielding effect, the transfer function between the substrate and the V_{tune} interconnect of the VCO, which is approximately 500 μm long (see Figure 7.3), is simulated with and without the ground shield. The simulation results are shown in Figure 7.4. Note that a ground plane enhances the isolation of the interconnect with approximately 30 dB in the frequency range of interest. Similar results are obtained in [8]. Remember

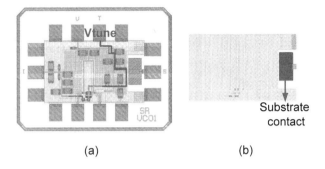

Figure 7.3 (a) The original layout and (b) the stripped layout.

that the impact of substrate noise does not only depend on the isolation but also on the sensitivity of the perturbed node.

After simplifying the layout, the remaining geometry of Figure 7.3(b) is imported into the EM simulator (HFSS) as explained in Chapter 2. In the HFSS environment, one lumped port is placed for each transistor. An additional lumped port is also placed at the substrate contact. This substrate contact is used to inject substrate noise into the substrate and hence replaces the switching digital circuitry. Consequently, the HFSS environment counts nine ports. This environment is simulated from DC up to 60 GHz with a minimum solved frequency of 50 MHz and a maximum error of the S-parameters of 0.01. The HFSS simulation takes about 37 minutes on an HP DL145 server.

7.4.2.2 Parasitic Extraction

An RC parasitic extraction is performed on the original layout with Calibre PEX [6]. The RC extraction takes 10 minutes on an HP Proliant DL145G1 platform.

7.4.2.3 Circuit Simulation

The resulting S-parameter box from the HFSS simulation and the RC mesh of the parasitic extraction are merged with the RF models of the devices into one netlist. In this example, it was not necessary to model the S-parameter box as a lumped network. A periodic AC analysis is performed on this netlist with SpectreRF [9]. This analysis takes 2 minutes on an HP Proliant DL145G1 platform. The resulting waveforms at the different nodes of the circuit reveal the dominant substrate noise coupling mechanisms.

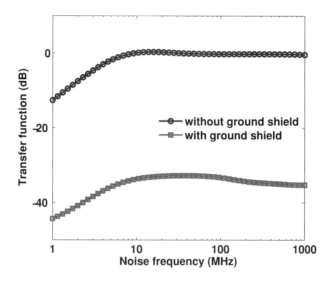

Figure 7.4   Influence of the ground shield on the transfer function between the substrate and the V_{tune} interconnect.

Figure 7.5 shows that the relative power of the sideband spurs match both for the simulation methodology of Chapter 6 and for the newly proposed simulation methodology. Similar conclusions about the substrate noise coupling mechanisms can also be drawn. However, in the simulation methodology of Chapter 6, the EM simulator needed to simulate all the details of the interconnects and the passive devices. As a consequence, the EM simulator had to solve the S-parameters at 28 ports instead of 9. The EM analysis took 7.5 hours on the same HP DL145 server. This means that our newly proposed methodology is 15 times faster. Furthermore, it is much easier to set up the EM environment. However, compared to the simulation methodology of the previous chapter an additional parasitic extraction needs to be performed. This does not form an obstacle as analog/RF and millimeter-wave designers routinely perform a parasitic extraction on their designs to estimate the capacitive loading of the circuit.

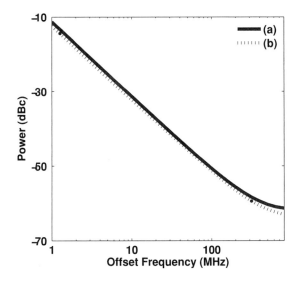

Figure 7.5 (a) Predicted sideband spurs with the newly proposed methodology. (b) Predicted sideband spurs with the simulation methodology of Chapter 6. Both methodologies obtain similar results.

7.4.3 Conclusion

The refined simulation methodology is used to predict the impact of substrate noise on a 48–53 GHz LC-VCO. The power of the spurs is predicted as accurately as in the simulation methodology of Section 7.4.2. However, the newly proposed methodology is 15 times faster. This clearly shows that the prior knowledge that substrate noise couples at low frequencies into the ground interconnect is applicable. Hence, the proposed methodology can be used to analyze the impact of substrate noise in larger analog/RF circuits and even on analog/RF systems. The next section uses this methodology to predict the impact of substrate noise on a 5 GHz wideband receiver.

7.5   IMPACT OF SUBSTRATE NOISE ON A DC TO 5 GHZ WIDEBAND RECEIVER

The main goal of the proposed methodology is to predict the impact of substrate noise on analog/RF systems in general. To that end, the general applicability of the proposed methodology is demonstrated on a DC to 5 GHz wideband receiver. This receiver is totally different than the millimeter-wave VCO of the previous section and thus the noise coupling mechanisms can be expected to be totally different. The receiver is integrated in a UMC 90 nm technology on a lightly doped substrate of 20 Ωcm. Note that this technology is different than the one used for the millimeter-wave VCO of the previous section. This shows that our proposed methodology is technology independent.

An overview of the complexity of the receiver is given first. The receiver achieves state-of-the-art performance and is therefore a representative candidate to study the impact of substrate noise. Next, the proposed methodology is applied and a simulation model is constructed in a way that is similar to the simulation setup of the LC-VCO. The different substrate coupling mechanisms are discussed and finally the simulations are validated with measurements.

7.5.1   Description of the Wideband Receiver

In order to predict and to understand the different substrate noise coupling mechanisms, it is mandatory for one to understand how the analog/RF system is built. Therefore this section briefly describes the different analog/RF circuits of which the wideband receiver consists. The receiver is realized in a 90 nm digital CMOS technology and offers an acceptable performance for a low area of 0.06 mm^2.

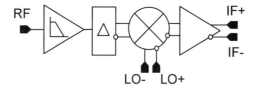

Figure 7.6   The receiver is a cascade of an LNA, a balun, a mixer, and an output buffer.

It consists of a cascade of an inductorless LNA, a balun, a downconversion mixer, and an output buffer (see Figure 7.6). Every circuit of the receiver is briefly described here.

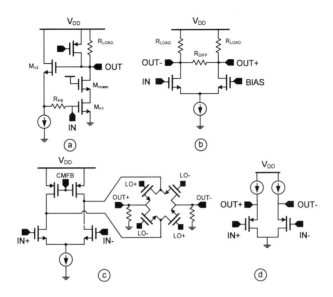

Figure 7.7 (a–d) Schematic of the different subcircuits.

• The LNA consists of a cascode amplifier with a source follower feedback path [10] [see Figure 7.7(a)]. Resistive degeneration in the feedback improves the linearity.

• The balun consists of a differential pair [see Figure 7.7(b)]. The differential pair is loaded with resistors whose value is chosen to maintain the high bandwidth at the price of gain.

• A mixer has been designed for low $1/f$-noise [see Figure 7.7(c)]. Very low current biases the switches while the transconductor stage draws the main current for high conversion gain. A common-mode feedback ensures DC stability of the mixer.

• The output buffer consists of a differential source follower [see Figure 7.7(d)] that drives the 50 Ω load of the measurement equipment while maintaining the linearity of the overall circuit.

All the circuits of the receiver do share the same ground. The different bias and supply lines are decoupled toward that ground with a large amount of MOS-based capacitors and MOM capacitors. A custom IO ring is made which provides sufficient ESD protection for bonding purposes.

The whole receiver provides DC-to-6 GHz amplification and downconversion. The receiver achieves an overall conversion gain of 33 dB (see Figure 7.8) and an NF down to 3.7 dB.

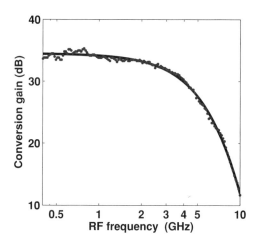

Figure 7.8   Measured and fitted conversion gain for a fixed IF frequency of 10 MHz.

The measured OIP3 is 2.4 dBm at 1 GHz and the 1 dB compression point is −41 dBm (see Figure 7.9). The linearity of this wideband receiver is studied in detail in [11].

Gaining insight into the different coupling mechanisms is complicated due to the large number of devices, the different feedback loops, and the frequency translation introduced by the mixer. Moreover, the nature of the coupling can be resistive, capacitive or both. When the designer lacks insight into the different substrate coupling mechanisms, it is very difficult —and close to impossible— to improve the immunity against substrate noise.

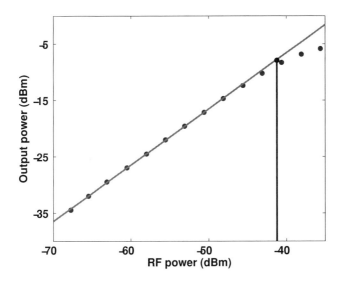

Figure 7.9   Measured 1 dB compression point at 1 GHz.

7.5.2   Simulation Setup

7.5.2.1   EM Simulation

The EM simulation setup is similar to the simulation setup of the LC-VCO. The layout of the receiver is also stripped until only the ground interconnect and the connections to the substrate remain.

The initial layout and the simplified layout are shown in Figure 7.10. Then, the layout is streamed in the HFSS environment.

The careful reader will notice that the VCO only contains NMOS devices and thus the EM simulation did not have to deal with the n-doped regions like, the n-well of the PMOS device and the MOS-based capacitors. As explained in Chapter 2, the capacitive behavior of the PN junction is modeled by inserting a silicon box with zero conductivity around the n-well. The width of the silicon box is determined by the PN junction capacitances as explained in Chapter 2.

The width of the depletion region at the lateral sides of the n-well is 200 nm and at the bottom side 1.1 μm. The depletion width at the lateral side of the n-well is

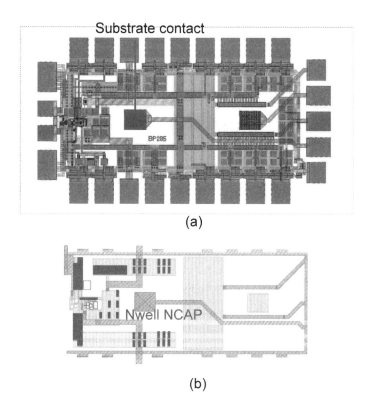

Figure 7.10   (a) Original layout. (b) Stripped layout.

much smaller than on the bottom side of the n-well because the p-well has a higher doping concentration than the lightly doped p-substrate. For small n-well regions substrate noise will couple more through the lateral sides of the n-well than through the bottom side of the n-well.

The wideband receiver counts a large number of n-wells. N-wells are a component of PMOS devices. They also shield MOS-based decoupling capacitors [see Figure 7.10(b)]. The n-wells of the decoupling capacitors are biased at the ground potential. An n-well biased at the ground potential still behaves capacitively [12]. In the EM environment, all the n-wells are surrounded by a silicon region which models the depletion region of the PN junction. Further, the HFSS environment is finalized with an airbox, the substrate, and the p-well similar to the HFSS environment of the VCO.

Afterwards ports are placed at the transistor locations and at the substrate contact. This HFSS environment is simulated from 100 MHz up to 2 GHz with a minimum solved frequency of 100 MHz and a maximum error of the S-parameters of 0.02. The HFSS simulation takes 1 hour and 33 minutes on a HP DL145 server which is very acceptable considering the complexity of the receiver.

7.5.2.2   Approximating the Substrate Transfer Function

To guarantee smooth convergence of the circuit analysis and to ensure good extrapolation of the S-parameters, the resulting S-parameter box is approximated by a first order RC network as explained in Section 7.2.1.4. Figure 7.11 shows the global transfer function between the substrate contact and the ground interconnect at the input transistor of the LNA [see Figure 7.7(a)]. The error made by this approximation is less than 1 dB. This RC network and the RC mesh of the parasitic extraction are merged with the models of the passive and active devices into one netlist.

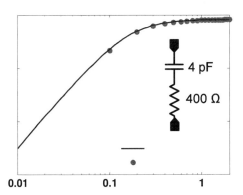

Figure 7.11   Overall transfer function between the substrate contact and the ground plane of the receiver.

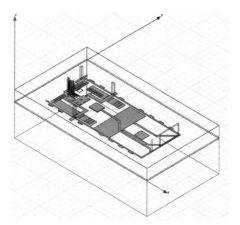

Figure 7.12   View of the HFSS environment.

7.5.3   Parasitic Extraction

The second simulation is an RC parasitic extraction. It is performed with Calibre PEX and takes 15 minutes on an HP Proliant DL145G1 server. The resulting RC meshes are automatically connected to the transistors in the simulation model.

7.5.4   Circuit Simulation

The third and final simulation is a circuit simulation. The models of the substrate and the interconnects are merged with the RF models of the different devices. Then mismatch effects are introduced as explained in Section 7.2.3 because they play an important role in the propagation of substrate noise through the receiver. Once the simulation model is complete, the simulation environment is set up and a transient simulation is performed as explained next.

7.5.4.1   Exciting the Receiver

The receiver is excited by a sine wave with a frequency that is chosen to be 1.04 GHz. Its power is chosen to be −50 dBm. The power of the sine wave is well below the 1 dB compression point of the receiver. This sine wave is downconverted to IF frequencies with an LO signal of 1.02 GHz and a power of -3 dBm. The IF frequency band of interest is centered around 20 MHz. The substrate can be excited independently by any signal of the designer's choice. Since the substrate can be considered to behave linearly, the power of the spurs scale linearly with the applied power. Note that the substrate behaves linearly but that substrate noise couples through a nonlinear receiver. In this case we excite the substrate with a square wave of 12.5 MHz because such a signal has a large number of harmonics which are visible in the IF band of interest. This square wave has an overall power of 0 dBm, a rise and fall time of 200 ns, and a duty cycle of 60%. The square wave is modeled based on measurements as a multitone of 8 tones.

7.5.4.2 Transient Simulation

A transient analysis is then performed using SpectreRF [9] on the simulation model of the receiver. The transient simulator is set up correctly; a small simulation step is mandatory to sample the high-frequency signal accurately. The simulator can only be stopped when the transient effects are faded away and when enough periods of the fundamental frequency are captured. The transient analysis takes about one hour on an HP Proliant DL145G1 platform. The resulting waveforms are discussed in the next subsection.

7.5.5 Revealing the Dominant Coupling Mechanism

First, this section builds up insight into the propagation mechanisms of substrate noise. Next, the coupling of substrate noise into the receiver and the generation of spurs will be discussed.

7.5.5.1 Propagation of Substrate Noise

Since substrate noise propagation is known to be mainly an electric effect, the propagation from the substrate contact toward the ground plane of the receiver is proportional to the electric field strength. Therefore, the substrate noise propagation can be visualized by plotting the simulated electric fields in the HFSS environment. Figure 7.13 shows the electric fields evaluated at the bottom of the p-well, the ground plane, and the n-wells at the minimum solved frequency of 100 MHz.

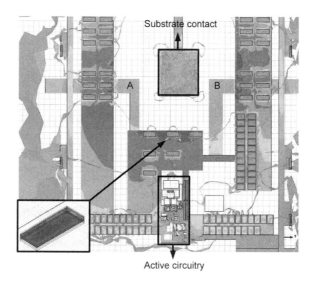

Figure 7.13   Simulated electric field in the ground interconnect, the p-wells, and the n-wells. (See color section.)

    From this figure, it can be seen that the magnitude of the electric fields in the ground plane increases in the neighborhood of the substrate contact. This seems to be counterintuitive. At low frequencies one would expect that substrate noise couples resistively into the receiver. In the region that is close to the substrate contact there is no resistive connection from the ground plane of the receiver to the substrate. The only resistive connections to the substrate are located near the transistors of the receiver and these are located farther away from the substrate contact. This means that substrate noise coupling is already dominated by capacitive effects at frequencies that are as low as 100 MHz. Substrate noise couples dominantly through the n-wells of the MOS-based decoupling capacitors to the ground interconnect and causes ground bounce. Remember that the n-wells of the decoupling capacitors are biased at the ground potential. Since it is a good design practice to decouple the bias lines as much as possible, there are many MOS-based decoupling capacitors and consequently a large amount of n-wells where substrate noise can couple through. It is also interesting to see that substrate noise couples more through the lateral side of the n-well than through the bottom side (see Figure 7.13). The n-well regions of the PMOS transistors are much smaller. Their effect is at least 30 dB lower and can safely be neglected.

As a consequence, substrate noise couples mainly capacitively into the ground plane. A part of the substrate noise is then drained through the two main connections (A and B in Figure 7.13) from the ground plane toward the PCB. Yet another part of the substrate noise reaches the transistors of the receiver, where it jeopardizes its performance.

In conclusion, the propagation of substrate noise is shown to be quite complex. The substrate flow is multidimensional as is shown in Figure 7.14. This figure shows the magnitude of the electric fields in the substrate. Small layout details such as n-wells clearly do influence the substrate noise propagation. In this case the n-wells even dominate the substrate noise propagation.

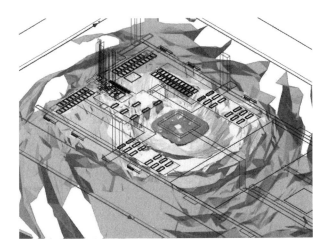

Figure 7.14   The current flow in the substrate is multidimensional. (See color section.)

7.5.5.2   Substrate Noise Coupling

When substrate noise couples into the receiver, spurs are created by different mechanisms. Therefore, it is interesting to first identify the origin of the spurs. In order to do this, it is important to rephrase that the receiver is excited at its input (RF) with a sine wave of 1.04 GHz ($=f_{RF}$) and its LO input (LO) with a sine wave of 1.02 GHz ($=f_{LO}$). Furthermore the substrate (sub) is excited with a square wave of 12.5 MHz ($=f_{sub}$). Figure 7.15 shows the simulated differential output spectrum of the receiver from DC up to 50 MHz.

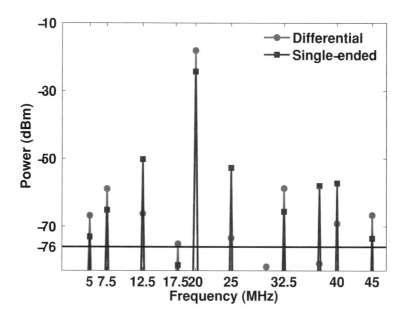

Figure 7.15   Simulated output spectrum of the receiver. The ● is the differential output. The □ is the single-ended output. The line at −76 dBm represents the measurement noise floor.

One can immediately identify the wanted tone located at 20 MHz $(f_{RF} - f_{LO})$ and its second harmonic located at 40 MHz $(2 \cdot f_{RF} - 2 \cdot f_{LO})$. The other — unwanted— spurs are created by the impact of substrate noise. Only the spurs with a differential output power higher than -76 dBm are considered because those spurs can be validated later on with measurements. The measurement noise floor is -76 dBm. Table 7.1 summarizes the frequency mixing process behind those unwanted spurs. The spurs can be classified into two groups. According to Table 7.1 the first group of spurs is the result of an upconversion with the RF tone of the receiver followed by a downconversion by the LO tone of the receiver. Hence the spurs lie in the IF band. To get more insight into the generation of these spurs, the single-ended simulated output is compared with the differential output spectrum (see Figure 7.15). For the first group of spurs the single-ended simulated power lies 6 dB lower than for the differential power. This means that the first group of spurs propagates through the circuit as a differential signal. Hence, they are very harmful because they cannot be distinguished from the desired signal ($f_{RF} - f_{LO}$). The second group of spurs contains all spurs that couple directly to the output without frequency translation. Here, the single-ended output power lies much higher than the differential output power. This indicates that those spurs propagate as a common-mode signal and can be removed by additional circuitry. Note that in the single-ended output spectrum of Figure 7.15, those spurs reach the highest power levels and will thus create the most distortion in the receiver.

Table 7.1

Spurs Generated by Substrate Noise from DC up to 50 MHz

Frequency (MHz)	Power (dBm)	Origin
Group 1		
17.5	-72	$f_{RF}\text{-}f_{LO}\text{-}3\mathrm{x}f_{sub}$
5	-66	$f_{RF}\text{-}f_{LO}\text{-}2\mathrm{x}f_{sub}$
7.5	-58	$f_{RF}\text{-}f_{LO}\text{-}f_{sub}$
32.5	-58	$f_{RF}\text{-}f_{LO}\text{+}f_{sub}$
45	-66	$f_{RF}\text{-}f_{LO}\text{+}2\mathrm{x}f_{sub}$
Group 2		
12.5	-67	f_{sub}
25	-71	$2\mathrm{x}f_{sub}$

At this point, the spurs are categorized and now the substrate coupling mechanisms will be revealed for each group of spurs separately. The nonlinearly induced spurs (group 1) are generated by capacitive coupling of substrate noise into the ground plane, causing ground bounce (see Figure 7.11). The source of the input transistor of the LNA is the most sensitive to ground bounce because any signal that couples into this device is converted to a differential signal by the balun and is amplified throughout the whole receiver chain. This explains the differential nature and the high power of the spurs.

The linear feedthrough group (group 2) of spurs is generated by a different coupling mechanism. Here, mismatches between the LO transistors are responsible for the unwanted spurs in the differential output spectrum. Those mismatches involve variations in the threshold voltage V_{th} and in the width w of the LO transistors but also phase mismatches in the external balun, which is used to generate the differential LO signal. Figure 7.16 compares the simulated output spectrum without mismatches with the one where mismatches are considered. Without mismatch, the common-mode spurs cancel each other perfectly [see Figure 7.16(b)]. When mismatch effects are included, this common-mode cancellation is incomplete, and thus the power of those spurs is the residual of an imperfect cancellation [see Figure 7.16(a)]. This shows that it is mandatory to include mismatch-induced effects in the simulation in order to be able to accurately characterize all the unwanted spurs.

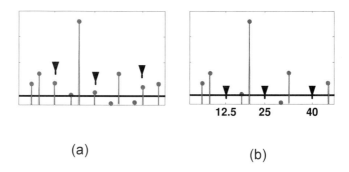

(a)   (b)

Figure 7.16   (a) Simulation with mismatches. (b) Simulation without mismatches. The lines at -76 dBm reflect the measurement noise floor.

7.5.6   Experimental Verification

Experimental verification is indispensable. Measurements on real-life ICs are indeed needed to build up the confidence in the proposed methodology. First, it will be demonstrated that the impact of the aggressor, which is a square wave of 12.5 MHz, is correctly modeled. Next, the measured output spectrum of the full receiver will be compared against the corresponding simulation.

7.5.6.1   Verification of the Properties of the Aggressor

The aggressor signal is a 12.5 MHz square wave that contains 0 dBm power. A perfect symmetrical square wave has only odd harmonics. In order to also create even harmonics the duty cycle is set to 60%. The signal is generated by a waveform generator and is measured with a oscilloscope to obtain both amplitude and phase information. The corresponding complex spectrum is modeled in the circuit simulator up to the eighth harmonic by eight current sources connected in parallel and terminated with a 50Ω resistor. The latter sets the output impedance level of the source. There is an excellent agreement between the time domain waveforms of the used model and the oscilloscope measurements (see Figure 7.17). Injecting such a square wave into the substrate contact that is located close to the receiver is therefore representative to verify the performed simulations.

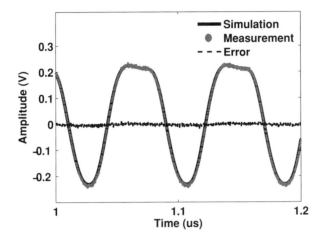

Figure 7.17 Measured versus modeled substrate signal. The line centered around zero reflects the error between the measured and the simulated waveforms.

7.5.6.2 Verification of the Output Spectrum of the Receiver

The IC of the wideband receiver (see Figure 7.18) is mounted on a PCB. The receiver is excited with the same signals as the ones used during simulation. The RF input is excited with a 1.02 GHz sine wave signal with a power of −50 dBm. This signal is provided by a continuous wave generator. The LO signal has a frequency of 1.04 GHz and −3 dBm power is also provided by a continuous wave generator. The substrate is excited by the square wave as shown above. The positive and the negative output of the receiver are combined with a balun. The differential output spectrum is measured from DC up to 50 MHz with a spectrum analyzer. As the spectrum analyzer only provides amplitude information, we can only compare the amplitudes of measured and simulated spectra. All the sources are synchronized

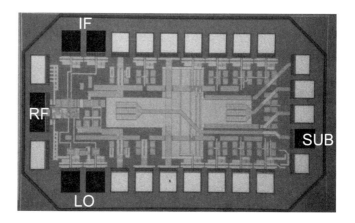

Figure 7.18   Chip photograph of the DC-to-5 GHz wideband receiver.

with the 10 MHz reference frequency of the spectrum analyzer to avoid frequency drift.

Figure 7.19 shows the measured and simulated differential output spectrum. The error in dB is marked on top of each spur. There is a very good agreement for the nonlinearly induced spurs (group 1). A good agreement is also obtained for the linear feed through a group of spurs (group 2) when an amount of mismatch is introduced that is equal to 1.3σ according to (7.2).

Therefore it can be concluded that the methodology successfully incorporates the dominant substrate-coupling mechanisms of the complete receiver.

7.6   CONCLUSIONS AND DISCUSSION

7.6.1   Conclusions

This chapter reduces the long simulation time that is inherent to the simulation methodology of Chapter 6 by taking advantage of the prior knowledge that substrate noise usually couples at low frequencies into the ground interconnect of an analog/RF circuit. The newly proposed methodology is at least one order of magnitude faster and can be used in an early stage of the design cycle. The main stronghold of this methodology is that it can easily be used by analog/RF designers. Indeed, this methodology combines the strength of three commercially available tools that

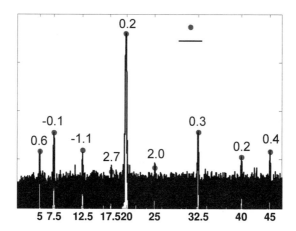

Figure 7.19 Measured versus simulated differential output spectrum. The error in decibels between simulations and measurements is marked on top of each spur.

are often used by analog designers: the EM simulator, the parasitic extractor, and the circuit simulator. Moreover, the designer does not have to worry about doping profiles. The capabilities of the methodology are demonstrated by means of two examples. First the substrate coupling mechanisms are revealed on a millimeter-wave LC-VCO. The second example is a wideband receiver. In just a few hours, the impact of substrate noise can be predicted on such a large circuit. As a proof of concept, all the performed simulations are successfully validated with measurements.

7.6.2 Discussion

The proposed methodology implicitly assumes that substrate noise couples at low frequencies into the ground interconnect. From our experience this is usually the case but one can think of specific cases where this assumption is not valid. Fortunately, the designer can involve any other interconnect in the EM simulation. For example, if the designer suspects that substrate noise couples dominantly into the input interconnect of the LNA, the designer can easily include this interconnect into the EM environment and check with the proposed methodology if the input interconnect of the LNA influences the substrate noise coupling mechanisms or not. Adding the input interconnect of the LNA into the EM environment will not affect the simulation time too much.

In any case, if substrate noise does not couple dominantly into the ground interconnect, there is still no need to involve all the interconnects of the analog/RF system in the EM environment. In the worst case, the designer will need a few try-outs based on the insight in the analog/RF system, before the dominant substrate noise coupling mechanisms can be revealed.

Further, the total number of ports can drastically be reduced. If the impedance of the ground interconnect is larger than 1Ω it is not necessary to place one port per transistor. One can assume that substrate noise will cause ground bounce. Hence, the bulkeffect can be neglected. One can then divide the entire ground interconnect in a number of regions. Each region is then assumed to be equipotential. The designer needs to place one port per region and hence the number of ports no longer depends on the number of transistors but on the number of user-defined regions. The number of regions is a trade-off between accuracy and complexity/simulation time. This has been tried on the example of the wideband receiver. As the receiver uses one ground plane, the transfer function from the substrate contact to the ground plane can be modeled with two ports only. Depending on the location of the port on the ground interconnect, this results in a worst case error of 10%.

References

[1] C. Soens, G. Van der Plas, P. Wambacq, S. Donnay, and M. Kuijk, "Performance degradation of LC-tank VCOs by impact of digital switching noise in lightly doped substrates," *IEEE Journal of Solid States*, Vol. 40, No. 7, July 2005, pp. 1472–1481.

[2] S. Bronckers, K. Scheir, G. Van der Plas, and Y. Rolain, "The impact of substrate noise on a 48-53GHz mm-wave LC-VCO," *Proc. IEEE Topical Meeting on Silicon Monolithic Integrated Circuits in RF Systems SiRF '09*, 2009, pp. 1–4.

[3] M. Felder and J. Ganger, "Analysis of ground-bounce induced substrate noise coupling in a low resistive bulk epitaxial process: design strategies to minimize noise effects on a mixed-signal chip," *IEEE Circuits and Systems II*, Vol. 46, No. 11, November 1999, pp. 1427–1436.

[4] S. Bronckers, C. Soens, G. Van der Plas, G. Vandersteen, and Y. Rolain, "Simulation methodology and experimental verification for the analysis of substrate noise on LC-VCO's," *Proc. Design, Automation & Test in Europe Conference & Exhibition, DATE '07*, 2007, pp. 1–6.

[5] R. Singh and S. Sali, "Modeling of electromagnetically coupled substrate noise in flash A/D converters," *IEEE Transactions on Electromagnetic Compatibility*, Vol. 45, No. 2, May 2003, pp. 459–468.

[6] Calibre PEX, http://www.mentor.com/products/ic_nanometer _design/bl_phy_design/calibre_xrc/.

[7] M. Pelgrom, A. Duinmaijer, and A. Welbers, "Matching properties of MOS transistors," *IEEE Journal of Solid States Circuits*, Vol. 24, No. 5, October 1989, pp. 1433–1439.

[8] A. Fanei, P. Pannier, J. Gaubert, M. Battista, and Y. Bachelet, "Experimental results and EM simulation of substrate noise in wideband low noise amplifier for UWB systems," *Proc. DTIS Design & Technology of Integrated Systems in Nanoscale Era International Conference*, 2007, pp. 192–195.

[9] SpectreRF, http://www.cadence.com/products/custom_ic/spectrerf.

[10] J. Borremans, P. Wambacq, and D. Linten, "An ESD-protected DC-to-6GHz 9.7mW LNA in 90nm Digital CMOS," *Proc. Digest of Technical Papers. IEEE International Solid-State Circuits Conference ISSCC 2007*, February 11–15, 2007, pp. 422–613.

[11] S. Bronckers, G. Vandersteen, J. Borremans, K. Vandermot, G. Van der Plas, and Y. Rolain, "Advanced nonlinearity analysis of a 6 GHz wideband receiver," *Proc. IEEE Instrumentation and Measurement Technology, IMTC 2008*, 2008, pp. 1340–1343.

[12] R. Vinella, G. Van der Plas, C. Soens, M. Rizzi, and B. Castagnolo, "Substrate noise isolation experiments in a 0.18 μm 1P6M triple-well CMOS process on a lightly doped substrate," *Proc. IEEE Instrumentation and Measurement Technology*, 2007, pp. 1–6.

Appendix A

Narrowband Frequency Modulation of LC-Tank VCOs

Low-frequency substrate noise that couples into an LC-VCO causes narrowband frequency modulated spurs [1]. This appendix derives an equation that can be used to calculate the power of the FM spurs. Consider a sinusoidal disturbance $V_{noise}(t)$ in the substrate:

$$V_{noise}(t) = A_{noise} \cdot cos(\omega_{noise}t) \qquad (A.1)$$

If $V_{noise}(t)$ is applied on a linear LTI system with one input and n outputs and with impulse response $h^i_{sub}(t)$, the response $V_i(t)$ will be, considering zero initial conditions:

$$V_i(t) = \int_0^t h^i_{sub}(\tau) \cdot V_{noise}(t - \tau)d\tau = A_{V_i}cos(\omega_{noise} + \phi_{V_i}) \qquad (A.2)$$

According to [2], the amplitude A_{V_i} and the phase ϕ_{V_i} of $V_i(t)$ can be expressed as follows:

$$A_{V_i} = |V_i(t)| = |H^i_{sub}(\omega_{noise})|A_{noise} \qquad (A.3)$$

$$\phi_{V_i} = \angle H^i_{sub}(\omega_{noise}) \qquad (A.4)$$

The equation describing the output of a VCO subject to frequency modulation by n perturbations is:

$$V_{out}(t) = A_{LO} \cdot (\omega_{LO}t + \sum_{i=1}^{n} 2\pi K_i \int_0^t V_i(t)) \qquad (A.5)$$

with $V_i(t) = 0, t < 0$. Here $K_i(V_{tune})$ ($\in \Re$) is the FM sensitivity function with unit (Hz/V) of node i of the VCO defined as:

$$K_i(V_{tune}) = \frac{\partial f_{LO}}{\partial V_i} \qquad (A.6)$$

Here f_{LO} is the local oscillator frequency, and V_{tune} is the tuning voltage of the VCO.

Substituting (A.2) in (A.5) gives:

$$V_{out}(t) = A_{LO} \cdot cos(\omega_{LO}t + \Delta\omega_{LO}t) \qquad (A.7)$$

with

$$\Delta\omega_{LO}t =$$

$$\sum_{i=1}^{n} 2\pi K_i \int_0^t A_{noise}|H^i_{SUB}(\omega_{noise})| \cdot cos(\omega_{noise}t + \angle H^i_{SUB}(\omega_{noise}))dt \quad (A.8)$$

When the frequency of $V_{out}(t)$ is integrated, the previous expression becomes:

$$V_{out}(t) = A_{LO} \cdot cos(\omega_{LO}t + \chi)$$
$$= A_{LO} \cdot cos(\omega_{LO}t) \cdot cos(+i) - sin(\omega_{LO}t) \cdot cos(\chi) \qquad (A.9)$$

with $\chi = \omega_{LO}t + \sum_{i=1}^{n} 2\pi K_i A_{noise}|H^i_{SUB}(\omega_{noise})| \cdot \frac{sin(\omega_{noise}t + \angle H^i_{SUB}(\omega_{noise}))}{\omega_{noise}}$

Since the disturbance in the substrate is assumed to be small compared to the local oscillator signal and assuming that

$\frac{\sum_{i=1}^{n} 2\pi K_i A_{noise} |H^i_{SUB}(\omega_{noise})| \cdot sin(\omega_{noise}t + \angle H^i_{SUB}(\omega_{noise}))}{\omega_{noise}}$ $\ll 1$, the following assumption can be made:

$$cos(\frac{\sum_{i=1}^{n} 2\pi K_i A_{noise} |H^i_{SUB}(\omega_{noise})|}{\omega_{noise}}$$
$$\cdot sin(\omega_{noise}t + \angle H^i_{SUB}(\omega_{noise})))) \approx 1 \tag{A.10}$$

and

$$sin(\frac{\sum_{i=1}^{n} 2\pi K_i A_{noise} |H^i_{SUB}(\omega_{noise})|}{\omega_{noise}}$$
$$\cdot sin(\omega_{noise}t + \angle H^i_{SUB}(\omega_{noise}))))$$
$$= \frac{\sum_{i=1}^{n} 2\pi K_i A_{noise} |H^i_{SUB}(\omega_{noise})|}{\omega_{noise}}$$
$$\cdot sin(\omega_{noise}t + \angle H^i_{SUB}(\omega_{noise}))) \tag{A.11}$$

Using Simpson's rule, this expression can be rewritten as:

$$V_{out}(t) = A_{LO}cos(\omega_{LO}t) +$$
$$\frac{\sum_{i=1}^{n} 2\pi K_i A_{noise} |H^i_{sub}(\omega_{noise})|}{2 \cdot \omega_{noise}} \cdot cos(\omega_{LO}t + \omega_{noise}t$$
$$+ \angle H^i_{sub}(\omega_{noise}) - cos(\omega_{LO}t - \omega_{noise}t + \angle H^i_{sub}(\omega_{noise}))) \tag{A.12}$$

Thus, at frequencies $f_{LO} \pm f_{noise}$ spurious tones will occur around the local oscillator signal with amplitude (divided by A_{LO}):

$$V_{out}(f_{LO} \pm f_{noise}) = \pm\frac{\sum_{i=1}^{n} K_i A_{noise} |H^i_{sub}(\omega_{noise})|}{2 \cdot f_{noise}} \tag{A.13}$$

This equation can be used to calculate the power of the sideband spurs caused by low frequency substrate noise coupling.

References

[1] S. Bronckers, G. Vandersteen, C. Soens, G. Van der Plas, and Y. Rolain, "Measurement and modeling of the sensitivity of LC-VCO's to substrate noise perturbations," *Proc. IEEE Instrumentation and Measurement Technology*, 2007, pp. 1–6.

[2] L. Rade and B. Westergren, *Beta Mathematics Handbook*,   Boca Raton, FL: CRC Press, 1997.

Appendix B

Port Conditions

Designers often use n-port networks in their design. For example, an amplifier is often described by a two-port network [1]. Such a two-port network has four terminals and four port variables (a voltage and a current at each port) (see Figure B.1). A pair of terminals is a port if the current that flows into one terminal is equal to the current that flows out of the other terminal. This current is determined by the network itself and the applied voltage difference at the terminals.

Figure B.1   General two-port network description.

A transistor can be described by a three-port network. Remember that this network describes in our case the interconnects and the substrate and not the transistor itself. The drain, gate, and bulk terminal of the transistor are referred to the source of the transistor. If the transistor is connected to this three-port network, its drain-source current I_{DS} is equal to:

$$I_{DS} = \frac{\mu C_{ox}}{2} \frac{W}{L} (V_{GS} - V_T)^2 \tag{B.1}$$

231

in the active region, with V_T proportional to V_{SB} and

$$I_{DS} = \frac{\mu C_{ox}}{2} \frac{W}{L} (2(V_{GS} - V_T)V_{DS} - V_{DS}^2) \tag{B.2}$$

in the triode region. Equations (B.1) and (B.2) point out that the current through a transistor with a given W/L is fully determined by the applied voltages V_{GS}, V_{DS}, and V_{BS}. In our case the W/L of the transistor is included in the RF model and the applied voltages are determined by the external bias voltages and currents and the voltage drop over the interconnects is determined by the n-port network. Moreover, the current that flows into the drain of the transistor leaves through the source of the transistor. Therefore our approach [2] satisfies the port conditions.

References

[1] P. R. Gray, P. J. Hurst, S. H. Lewis, and R. Meyer, *Analysis and Design of Analog Integrated Circuits*, New York: John Wiley & Sons, 2001.

[2] S. Bronckers, G. Vandersteen, J. Borremans, K. Vandermot, G. Van der Plas, and Y. Rolain, "Advanced nonlinearity analysis of a 6 GHz wideband receiver," *Proc. IEEE Instrumentation and Measurement Technology, IMTC 2008*, 2008, pp. 1340–1343.

List of Acronyms

AC alternating current

AM amplitude modulation .

$CMOS$ complementary metal oxide semiconductor

DC direct current

$DIVA$ dynamic implementation verification architecture

DNW deep n-well

DRC design rule check

ESD electrostatic discharge

FDD frequency division duplex

FDM finite difference method

FEM finite element method

FM frequency modulation

GSG ground-signal-ground

$HFSS$ high-frequency structural simulator

IF intermediate frequency

IO input-output

K_{VCO} VCO gain

LNA low-noise amplifier

LO local oscillator

MOM metal-oxide-metal

MOS metal oxide semiconductor

NF noise figure

$OIP3$ output referred third-order modulation product

PA power amplifier

PAC periodic AC

PCB printed circuit board

PEX parasitic extraction

PLL phase locked loop

PPA prepower amplifier

PSS periodic steady state

RF radio frequency

$S-parameters$ scattering parameters

SMA subminiature version A

SMD surface mounted device

SNA substrate noise analyst

SNR signal-to-noise ratio

SoC system on a chip

STI shallow trench isolation

TF transfer function

TSV through silicon VIA

$UMTS$ universal mobile telecommunications system

VCO voltage controlled oscillator

VNA vector network analyzer

$WLAN$ wireless local area network

About the Authors

Stephane Bronckers graduated as an electrical engineer in electronics and information processing in 2005 (master of science engineering) at the Vrije Universiteit Brussel (VUB). In 2006 he joined the Fundamental Electricity and Instrumentation (ELEC) Department at the VUB as a research assistant supported by the Institute for the Promotion of Innovation Through Science and Technology in Flanders (IWT-Vlaanderen). He worked at the Wireless Group of IMEC (Interuniversity Microelectronics Center) in Leuven, Belgium where he obtained his Ph.D. degree. His research focused on developing a methodology to improve the electromagnetic immunity of analog integrated RF circuits. In 2009, he joined the EMC Group of Laborelec, which is part of the GDF-Suez group.

Geert Van der Plas obtained an M.Sc. and a Ph.D. from the Katholieke Universiteit Leuven, Belgium, in 1992 and 2001, respectively. From 1992 to 2002, he was a research assistant with the ESAT-MICAS Laboratory of the Katholieke Universiteit Leuven. Since 2003, he has been with the Interuniversity Microelectronics Center (IMEC), Belgium, where he works as a principal scientist on energy efficient data converters, low-power scalable radios, noise coupling, signal integrity, and design technology for 3D integration. He is the author and coauthor of over 100 papers in journals and conference proceedings and serves on the technical program committee of the Symposium on VLSI Circuits.

Gerd Vandersteen received an electrical engineering degree from the Vrije Universiteit Brussel (VUB), Brussels, Belgium, in 1991. In 1997, Dr. Vandersteen received a Ph.D. in electrical engineering from the Vrije Universiteit Brussel/ELEC. During his postdoc he worked at the Interuniversity Microelectronics Center

(IMEC) as a principal scientist in the Wireless Group with a focus on the modeling, measurement, and simulation of electronic circuits in state-of-the-art silicon technologies. This research was in collaboration with the Vrije Universiteit Brussels. Since 2008 Dr. Vandersteen has been a professor at the Vrije Universiteit Brussels/ELEC, teaching the subject of measuring, modeling, and analysis of complex linear and nonlinear systems.

Yves Rolain is a burgerlijk ingenieur (EE-1984), has a master in computer sciences (1986), and a doctor in applied sciences (1993), all from the Vrije Universiteit Brussel (VUB). He was elected an IEEE Fellow in 2005 and is the recipient of the 2005 IEEE Instrumentation and Measurement Society Award. Dr. Rolain is currently a professor working for the Electrical Measurement Department (ELEC) of the engineering faculty of the VUB, where he teaches high-frequency electronics and electrical measurement techniques. His main research interests are nonlinear microwave measurement techniques, applied digital signal processing, and parameter estimation/system identification.

Index

3D SoC, 179
3D stacking, 179–190

Active device, 80

Balun, 208
Body transconductance, 81
Bond wire, 66, 128, 140, 154, 159
Buffer, 208
Bulk effect, 80, 92, 106, 167, 224
Bulk resistance, 100

Capacitor
    MIM, 97
    MOM, 209
    MOS, 215
    Parasitic, 159, 175
    SMD, 121, 154
Circuit simulation, 86, 154, 168, 183, 200, 214
CMOS
    Triple well, 97
    Twin well, 20
Conductivity, 30
Contact-to-contact resistance, 9–13
Coupling
    Capacitive, 24, 47, 52, 59, 135, 159, 183, 215
    Inductive, 127, 135
    Resistive, 23, 47, 92, 127, 134, 158
Cross-talk, see Coupling
Current mirror, 178

Depletion region, 30, 199, 211
Dice, 143
Die seal, 143
Die-on-die method, 181
Differential, 217
Diode, 97
Direct coupling, 112
Doped region
    Conductivity, 30
    Junction capacitance, 30
Doping profile, 20, 80, 154

Electrical field distribution, 33, 46, 104, 202, 215
EM simulation, 28, 44, 83, 163, 183, 198, 211

FDM, 13–16, 151
FEM, 17–19, 151
    Accuracy, 18
    Basis functions, 18
    Mesh, 17
    Ports, 18
Field solver, see EM simulations
Filter, 116
Flip chip, 66

Ground bounce, 80, 92, 106, 127, 158, 217, 224
Ground interconnect, 96, 105, 174
Guard ring, 40–42
    Floating, 61
    Grounded, 61
    N-well isolation, 50
    P^+ guard ring shielding, 52
    P-well block isolation, 47
    Triple well shielding, 56

HFSS, 33

Immunity, *see* Substrate noise immunity
Impedance function, 154
Inductor, 140, 175
    Parasitic, 131
Insulator
    Relative permittivity, 30

LC-VCO, *see* Voltage controlled oscillator
Limiter, 115, 168, 179
Low noise amplifier, 116, 208
Lumped network, 95, 152

Magnetic coupling, *see* Coupling
Maxwell equations, 8
Metal
    Conductivity, 30
Millimeter wave, 166
Mismatch, 200, 217
Mixer, 171, 208
Modeling
    Bond wire, 155
    Capacitor, 154
    Interconnect, 84, 154
    Interconnects, 215
    Substrate, 8, 17
    Trace, 155
Modulation
    AM, 111, 122, 159, 175
    FM, 111, 122, 158, 168, 174, 183
MOM capacitor, 209
MOS-based capacitor, 215

N-port, 61
N-well isolation, 50–52

On-wafer probes, 99, 170
Order reduction modeling, 200

P^+ guard ring shielding, 52–56, 65
P-well block isolation, 43, 47–50
Parasitic extraction, 142, 200, 214
Passive isolation structure, *see* Guard ring
Phase locked loop, 136
PN junction, 30, 50, 102, 199, 211

Port, 42, 85
Port condition, 86
Power amplifier, 112, 133
Prepower amplifier (PPA), *see* Power amplifier
Printed circuit board, 118, 183
    Decoupling capacitor, 119, 121, 130, 154
    Trace, 155

Relative permittivity, 30
Resonance, 154, 170
RF model, 80

S-parameters, 23, 46, 104, 199
Scaling, *see* Technology scaling
Sensitivity factor, 116
Sensitivity function, 116–118, 171
    AM, 128
    FM, 123, 168
Sheet resistance, 174
Sideband spurs, 112, 138, 156, 167
Sizeable guard ring, 67
Skin effect, 16
Substrate
    Cutoff frequency, 15, 49
    Lightly doped, 3
    Relaxation time, 15
    Transfer function, 199, 213
Substrate contact, 3, 19, 42, 97, 138, 161, 181,
        220
Substrate noise, 2
    Propagation, 7–34
Substrate noise immunity, 174–180
SubstrateStorm, 19
Switching activity, 112
System in a cube, 179
System on a chip, 1

Technology scaling, 174
Tooling, 7
Transconductance, 81
Transfer function, 96, 115
Transistor, *see* Active device
Triple well shielding, 32, 56–97
Twin well, 99

Varactor, 176

Voltage controlled oscillator, 111, 119, 156,
         166, 203
    Switched varactor, 136, 176

Wideband receiver, 208

Y-parameters, 73, 104, 199

Recent Titles in the Artech House Microwave Library

Active Filters for Integrated-Circuit Applications, Fred H. Irons

Advanced Techniques in RF Power Amplifier Design, Steve C. Cripps

Automated Smith Chart, Version 4.0: Software and User's Manual,
   Leonard M. Schwab

Behavioral Modeling of Nonlinear RF and Microwave Devices,
   Thomas R. Turlington

Broadband Microwave Amplifiers, Bal S. Virdee, Avtar S. Virdee, and
   Ben Y. Banyamin

Computer-Aided Analysis of Nonlinear Microwave Circuits,
   Paulo J. C. Rodrigues

Designing Bipolar Transistor Radio Frequency Integrated Circuits,
   Allen A. Sweet

Design of FET Frequency Multipliers and Harmonic Oscillators,
   Edmar Camargo

Design of Linear RF Outphasing Power Amplifiers, Xuejun Zhang,
   Lawrence E. Larson, and Peter M. Asbeck

Design Methodology for RF CMOS Phase Locked Loops,
   Carlos Quemada, Guillermo Bistué, and Iñigo Adin

*Design of RF and Microwave Amplifiers and Oscillators, Second
   Edition*, Pieter L. D. Abrie

Digital Filter Design Solutions, Jolyon M. De Freitas

*Discrete Oscillator Design Linear, Nonlinear, Transient, and Noise
   Domains*, Randall W. Rhea

Distortion in RF Power Amplifiers, Joel Vuolevi and Timo Rahkonen

*EMPLAN: Electromagnetic Analysis of Printed Structures in Planarly
   Layered Media, Software and User's Manual*, Noyan Kinayman
   and M. I. Aksun

Essentials of RF and Microwave Grounding, Eric Holzman

FAST: Fast Amplifier Synthesis Tool—Software and User's Guide, Dale D. Henkes

Feedforward Linear Power Amplifiers, Nick Pothecary

Foundations of Oscillator Circuit Design, Guillermo Gonzalez

Fundamentals of Nonlinear Behavioral Modeling for RF and Microwave Design, John Wood and David E. Root, editors

Generalized Filter Design by Computer Optimization, Djuradj Budimir

High-Linearity RF Amplifier Design, Peter B. Kenington

High-Speed Circuit Board Signal Integrity, Stephen C. Thierauf

Intermodulation Distortion in Microwave and Wireless Circuits, José Carlos Pedro and Nuno Borges Carvalho

Introduction to Modeling HBTs, Matthias Rudolph

Lumped Elements for RF and Microwave Circuits, Inder Bahl

Lumped Element Quadrature Hybrids, David Andrews

Microwave Circuit Modeling Using Electromagnetic Field Simulation, Daniel G. Swanson, Jr. and Wolfgang J. R. Hoefer

Microwave Component Mechanics, Harri Eskelinen and Pekka Eskelinen

Microwave Differential Circuit Design Using Mixed-Mode S-Parameters, William R. Eisenstadt, Robert Stengel, and Bruce M. Thompson

Microwave Engineers' Handbook, Two Volumes, Theodore Saad, editor

Microwave Filters, Impedance-Matching Networks, and Coupling Structures, George L. Matthaei, Leo Young, and E.M.T. Jones

Microwave Materials and Fabrication Techniques, Second Edition, Thomas S. Laverghetta

Microwave Mixers, Second Edition, Stephen A. Maas

Microwave Radio Transmission Design Guide, Second Edition, Trevor Manning

Microwaves and Wireless Simplified, Third Edition,
  Thomas S. Laverghetta

Modern Microwave Circuits, Noyan Kinayman and M. I. Aksun

Modern Microwave Measurements and Techniques, Second Edition,
  Thomas S. Laverghetta

Neural Networks for RF and Microwave Design, Q. J. Zhang and
  K. C. Gupta

Noise in Linear and Nonlinear Circuits, Stephen A. Maas

Nonlinear Microwave and RF Circuits, Second Edition,
  Stephen A. Maas

*QMATCH: Lumped-Element Impedance Matching, Software and
  User's Guide,* Pieter L. D. Abrie

Practical Analog and Digital Filter Design, Les Thede

Practical Microstrip Design and Applications, Günter Kompa

*Practical RF Circuit Design for Modern Wireless Systems, Volume I:
  Passive Circuits and Systems,* Les Besser and Rowan Gilmore

*Practical RF Circuit Design for Modern Wireless Systems, Volume II:
  Active Circuits and Systems,* Rowan Gilmore and Les Besser

*Production Testing of RF and System-on-a-Chip Devices for Wireless
  Communications,* Keith B. Schaub and Joe Kelly

Radio Frequency Integrated Circuit Design, Second Edition,
  John W. M. Rogers and Calvin Plett

RF Bulk Acoustic Wave Filters for Communications,
  Ken-ya Hashimoto

RF Design Guide: Systems, Circuits, and Equations, Peter Vizmuller

RF Measurements of Die and Packages, Scott A. Wartenberg

The RF and Microwave Circuit Design Handbook, Stephen A. Maas

RF and Microwave Coupled-Line Circuits, Rajesh Mongia, Inder Bahl,
  and Prakash Bhartia

RF and Microwave Oscillator Design, Michal Odyniec, editor

RF Power Amplifiers for Wireless Communications, Second Edition, Steve C. Cripps

RF Systems, Components, and Circuits Handbook, Ferril A. Losee

The Six-Port Technique with Microwave and Wireless Applications, Fadhel M. Ghannouchi and Abbas Mohammadi

Solid-State Microwave High-Power Amplifiers, Franco Sechi and Marina Bujatti

Stability Analysis of Nonlinear Microwave Circuits, Almudena Suárez and Raymond Quéré

Substrate Noise Coupling in Analog/RF Circuits, Stephane Bronckers, Geert Van der Plas, Gerd Vandersteen, and Yves Rolain

System-in-Package RF Design and Applications, Michael P. Gaynor

TRAVIS 2.0: Transmission Line Visualization Software and User's Guide, Version 2.0, Robert G. Kaires and Barton T. Hickman

Understanding Microwave Heating Cavities, Tse V. Chow Ting Chan and Howard C. Reader

For further information on these and other Artech House titles, including previously considered out-of-print books now available through our In-Print- Forever® (IPF®) program, contact:

Artech House Publishers
685 Canton Street
Norwood, MA 02062
Phone: 781-769-9750
Fax: 781-769-6334
e-mail: artech@artechhouse.com
artech-uk@artechhouse.com

Artech House Books
16 Sussex Street
London SW1V 4RW UK
Phone: +44 (0)20 7596 8750
Fax: +44 (0)20 7630 0166
e-mail:

Find us on the World Wide Web at: www.artechhouse.com